Quantum Hall Effect

The quantum Hall effect (QHE) is a fundamental phenomenon that occurs in a two-dimensional electron gas (2DEG) at low temperature and in the presence of a strong magnetic field. It has various applications in fields such as metrology and topological quantum computers. It also provides an extremely precise and independent determination of the fine-structure constant—a quantity of fundamental importance in quantum electrodynamics.

This book attempts to present concepts of QHE to undergraduate and graduate students, post-doctoral researchers, and teachers taking advanced courses on condensed matter physics. The author has integrated all important concepts of QHE such as graphene, the connection between topology and condensed matter physics, and prospects of next-generation storage devices based on the manipulation of spins (spintronic) and presented them in a lucid manner. It offers the advantage of providing a pedagogical presentation to help students with some intermediate steps in derivation.

The book starts with an introduction to the experimental discovery of the QHE that segues into the basics of 2DEG in a magnetic field. The physics of the Landau levels, their properties, and their relevance to the integer QHE are discussed. The importance of conduction and its connection to topological insulators is also emphasised. At a pedagogical level, concepts such as linear response theory, Kubo formula, and topological invariance are explained and their relations to the understanding of QHE, graphene, its symmetries, and its relevance as a quantum Hall insulator are also covered. It ends with an explanation of the role of interparticle interactions to explain fractional QHE with the help of topics such as the Laughlin wave function, fractional charge and statistics, and non-abelian anyons.

Saurabh Basu is Professor in the Department of Physics at IIT Guwahati. His research interests include different fields of theoretical condensed matter physics with a focus on topological materials, higher order topological insulators, ultracold physics, non-Hermitian systems, charge and thermal transport in mesoscale and nanoscale devices, Floquet dynamics, critical phenomena and fractals, and Skyrmions, among others. He has also developed online courses on condensed matter physics, quantum mechanics, superconductivity, and numerical methods which are hosted on SWAYAM, a popular online free e-learning platform initiated by the Government of India. He has also written a few popular science articles for the general audience, which have appeared in different Indian languages.

Quantum Hall Effect

The First Topological Insulator

Saurabh Basu

CAMBRIDGE
UNIVERSITY PRESS

CAMBRIDGE
UNIVERSITY PRESS

Shaftesbury Road, Cambridge CB2 8EA, United Kingdom

One Liberty Plaza, 20th Floor, New York, NY 10006, USA

477 Williamstown Road, Port Melbourne, VIC 3207, Australia

314–321, 3rd Floor, Plot 3, Splendor Forum, Jasola District Centre, New Delhi – 110025, India

103 Penang Road, #05–06/07, Visioncrest Commercial, Singapore 238467

Cambridge University Press is part of Cambridge University Press & Assessment, a department of the University of Cambridge.

We share the University's mission to contribute to society through the pursuit of education, learning and research at the highest international levels of excellence.

www.cambridge.org
Information on this title: www.cambridge.org/9781316511756

First published 2024

Printed in India by Avantika Printers Pvt. Ltd.

A catalogue record for this publication is available from the British Library

Library of Congress Cataloging-in-Publication Data

Names: Basu, Saurabh, author.
Title: Quantum Hall effect : the first topological insulator / Saurabh Basu.
Description: Cambridge, United Kingdom ; New York, NY : Cambridge University Press, 2024. | Includes bibliographical references and index.
Identifiers: LCCN 2024011919 (print) | LCCN 2024011920 (ebook) | ISBN 9781316511756 (hardback) | ISBN 9781009053778 (ebook)
Subjects: LCSH: Quantum Hall effect. | Electron gas.
Classification: LCC QC612.H3 B37 2024 (print) | LCC QC612.H3 (ebook) | DDC 537.6/226--dc23/eng/20240409
LC record available at https://lccn.loc.gov/2024011919
LC ebook record available at https://lccn.loc.gov/2024011920

ISBN 978-1-316-51175-6 Hardback

To my family...

Contents

Foreword

Traditionally the different states of matter are described by symmetries that are broken. Typical situations include the freezing of a liquid, which breaks the translational symmetry that the fluid possessed, and the onset of magnetism, where the rotational symmetry is broken by the ordering of the individual magnetic moment vectors. In the early eighties of the previous century a completely new organizational principle of quantum matter was introduced following the discovery of the quantum Hall effect. The robustness of the quantum Hall state was a forerunner of the variety of topologically protected states that forms a large fraction of the condensed matter physics and material science literature at present.

Given the rapid strides that this field has made in the last two decades, it is almost imperative that it should become a part of the senior undergraduate curriculum. This necessitates the existence of a textbook that can address these somewhat esoteric topics at a level which is understandable to those who have not yet decided to specialize in this particular field but very well could, if given a proper exposition. This is a rather difficult task for the author of a textbook of a contemporary topic, and this is where the present book is immensely successful.

I am not a specialist in this subject by any means and found the book to be a comprehensive introduction to the area. I am sure the senior undergraduates and the beginning graduate students will benefit immensely from the book.

J.K. Bhattacharjee

Jayanta K. Bhattacharjee
School of Physical Sciences,
Indian Association for the Cultivation of Science
Jadavpur, Kolkata

Preface

It is somewhat implicit that the readers are familiar with the first course on *solid state physics*, which mainly deals with electronic systems and teaches us how to distinguish between different forms of matter, such as metals, semiconductors and insulators. An elementary treatise on band structure is introduced in this regard, and in most cases, interacting phenomena, such as magnetism and superconductivity, are taught. The readers are encouraged to look at the classic texts on solid state physics, such as the ones by Kittel, Ashcroft and Mermin.

As a second course, or an advanced course on the subject, more in-depth study of condensed matter physics and its applications to the physical properties of various materials have found a place in the undergraduate curricula for a century or even more. The perspective on teaching the subject has remained unchanged during this period of time. However, the recent developments over the last few decades require a new perspective on teaching and learning about the subject. Quantum Hall effect is one such discovery that has influenced the way condensed matter physics is taught to undergraduate students. The role of topology in condensed matter systems and the fashion in which it is interwoven with the physical observables need to be understood for deeper appreciation of the subject. Thus, to have a quintessential presentation for the undergraduate students, in this book, we have addressed selected topics on the quantum Hall effect, and its close cousin, namely topology, that should comprehensively contribute to the learning of the topics and concepts that have emerged in the not-so-distant past. In this book, we focus on the transport properties of two-dimensional (2D) electronic systems and solely on the role of a constant magnetic field perpendicular to the plane of a electron gas. This brings us to the topic of quantum Hall effect, which is one of the main verticals of the book. The origin of the Landau levels and the passage of the Hall current through edge modes are also discussed. The latter establishes a quantum Hall sample to be the first example of a topological insulator. Hence, our subsequent focus is on the subject topology and its application to quantum Hall systems and in general to condensed matter physics. Introducing the subject from a formal standpoint, we discuss the band structure and topological invariants in 1D. In particular, we talk about the Su–Schrieffer–Heeger and the Kitaev models in 1D, which, apart from being a possible realization for a polyacetylene molecule and a tight-binding chain with p-wave superconducting correlations, respectively, have emerged as a paradigmatic tool to study topology in 1D tight-binding systems. We have further discussed a quasi-1D system, namely the Creutz ladder. Owing to its existence intermediate to 1D and 2D, and the inability to put in the conventional classification of the topological insulators, the model, although quite interesting, received less attention. It is important to mention that these models may not be directly related with quantum Hall effect owing to time reversal symmetry being intact. Nevertheless, they are important to understand the topological ideas that are under the lens as well.

Having discussed quantum Hall effect in a 2D electron gas, it is of topical interest to discuss the corresponding scenario on a 2D crystal lattice. Or in a different sense, one may wish to distinguish between the behaviour of the relativistic and the non-relativistic electrons in two dimensions in the presence of a transverse magnetic field. In this context graphene is important, and it is amply elaborated in the text for its reason to be important. The formation of the Landau levels, which is central to the understanding of quantum Hall effect, is discussed, and, quite interestingly, owing to the large spacing between consecutive Landau levels in graphene, one should be able to observe the quantum Hall effect at room temperature. Being able to experimentally measure quantum effects in the classical regime is indeed a significant discovery.

We also deliberate upon the possibility pointed out by Haldane whether magnetic field is indispensable for realizing quantum Hall effect. A related topic that had ignited interest and debate is whether graphene can become a topological insulator. It turns out that being able to break the time reversal symmetry is more fundamental than the presence of an external field. This brings us to the topic of anomalous quantum Hall effect in graphene, which also implies that upon suitably tweaking the Hamiltonian, graphene can become a topological insulator. Addition of the spin of the electrons to the ongoing discussion emerged as a unique possibility to yield another version of the topological insulator, namely the quantum spin Hall insulator, which may lie at the heart of the next-generation spintronic devices.

Thereafter, a crisp introduction to the fractional quantum Hall effect is included. It comprises the discussion of Laughlin states, composite fermions and the hierarchy scenario, which will benefit the students in understanding the role of electronic interactions resulting in fractionally quantized Hall plateaus. We also briefly discuss the particle statistics in 2D, known as the braiding statistics, and touch upon how it aids in solving the riddle of even denominator fractions observed in experiments. A very brief introduction to the fractional quantum Hall effect in graphene has been included at the end.

All the while during the course of the book, we have included rigorous mathematical derivations wherever required, presented experimental details to connect with the ongoing discussions and tried to be as lucid as possible in our presentation of topics and concepts. A whole lot of schematic diagrams are presented for clarity as well. We hope that the students gain from the essence of this book, and it aids their understanding of both the topical and the traditional condensed matter physics. We shall be available and happy to answer queries, provide clarifications to students and researchers, and welcome comments for improvement.

Acknowledgements

It is a proud privilege to acknowledge a lot of people who have actively or passively contributed in bringing up the book in the current form. First and foremost, we owe a lot to a number of current and past graduate students of the Department of Physics, IIT Guwahati, such as Ms Shilpi Roy, Mr Sayan Mandal, Mr Dipendu Halder, Mr Koustav Roy and Ms Srijata Lahiri, Ms. Shreya Debnath, Dr Sudin Ganguly, Dr Sk. Noor Nabi, Dr Priyadarshini Kapri and Dr Priyanka Sinha. I am particularly thankful to Shilpi, Sayan and Srijata for contributing in several ways during my preparation of the manuscript. It is also a privilege to acknowledge a number of people who have contributed directly or indirectly to the preparation of the book. They are Prof. Gaurav Dar (BITS Pilani, Goa campus), Prof. A Perumal (HoD, Physics, IIT Guwahati), Dr. Kuntal Bhattacharya (Post-doctoral fellow, IIT Guwahati), Prof. Dilip Pal (IIT Guwahati), Prof. Krishnendu Sengupta (IACS, Kolkata), Prof. Arindam Ghosh (IISc Bangalore), Prof. Tapan Mishra (NISER, Bhubaneswar), Prof. Pankaj Mishra (IIT Guwahati), Prof. Sumiran Pujari (IIT Bombay), Prof. Siddharth Lal (IISER Kolkata) and others. I also thank my daughters Shreya and Shreemoyee for being big critics of me writing this book.

1

Quantum Hall Effect

1.1 Introduction

The date of discovery of the quantum Hall effect (QHE) is known pretty accurately. It occurred at 2:00 a.m. on 5 February 1980 at the high magnetic lab in Grenoble, France (see Fig. 1.1). There was an ongoing research on the transport properties of silicon field-effect transistors (FETs). The main motive was to improve the mobility of these FET devices. The devices that were provided by Dorda and Pepper allowed direct measurement of the resistivity tensor. The system is a highly degenerate two-dimensional (2D) electron gas contained in the inversion layer of a metal oxide semiconductor field effect transistor (MOSFET) operated at low temperatures and strong magnetic fields. The original notes appear in Fig. 1.1, where it is clearly stated that the Hall resistivity involves universal constants and hence signals towards the involvement of a very fundamental phenomenon.

In the classical version of the phenomenon discovered by E. Hall in 1879, just over a hundred years before the discovery of its quantum analogue, one may consider a sample with a planar geometry so as to restrict the carriers to move in a 2D plane. Next, turn on a bias voltage so that a current flows in one of the longitudinal directions and a strong magnetic field perpendicular to the plane of the gas (see Fig. 1.2). Because of the Lorentz force, the carriers drift towards a direction transverse to the direction of the current flowing in the sample. At equilibrium, a voltage develops in the transverse direction, which is known as the Hall voltage. The Hall resistivity, R, defined as the Hall voltage divided by the longitudinal current, is found to linearly depend on the magnetic field, B, and inversely on the carrier density, n, through $R = \frac{B}{nq}$ (q is the charge). A related

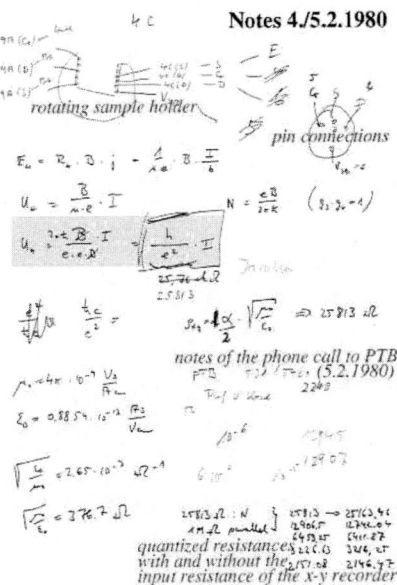

FIGURE 1.1.   Copy of the original notes by K.v. Klitzing on the discovery of the quantum Hall effect. It documents that the Hall resistance ($\frac{U_H}{I}$) involves a fundamental constant h/e^2 (taken from Ref. [60]).

and possibly more familiar quantity is the Hall coefficient, denoted by $R_H = R/B$, which via its sign yields information on the type of the majority carriers, that is, whether they are electrons or holes.

At very low temperature or at very high values of the magnetic field (or at both), the resistivity of the sample assumes quantized values of the form $\rho_{xy} = \frac{h}{ne^2}$. Initially, n was found to be an integer with extraordinary precession (one part in $\sim 10^8$). This is shown in Fig. 1.3. The quantization of the Hall resistivity yields the name *quantum* (or quantized) Hall effect, which we refer to as QHE throughout the book.

Klaus von Klitzing and his co-workers [60, 61], while measuring the electrical transport properties of planar systems formed at the interface of two different semiconducting samples in the strong magnetic field facility at Grenoble, France, noted that the Hall resistivity is quantized in units of h/e^2 as a function of the external magnetic field. The flatness of the plateaus occurring at integer or fractional values of h/e^2 has an unprecedented precession and is independent of the geometry of the sample (as long as it is 2D), density of the charge carriers and its purity. The accuracy of the quantization aids in fixing the unit of resistance, namely $h/e^2 = 25.813$ kΩ, also called the Klitzing constant. Thus, among other significant properties of QHE that we shall be discussing in due course, an experiment performed at a macroscopic scale that can be used for metrology

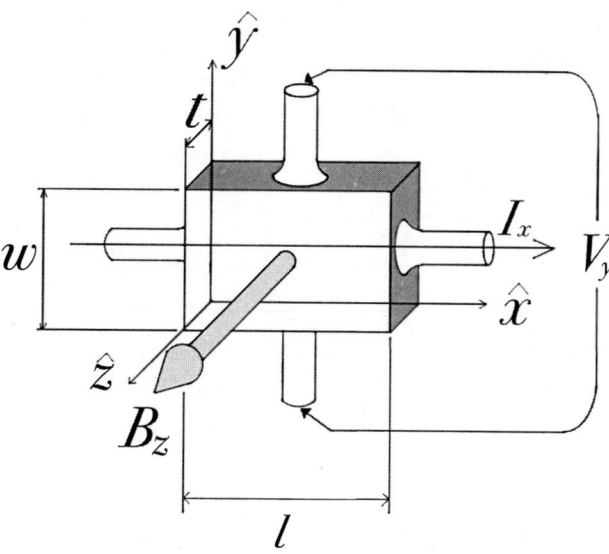

FIGURE 1.2. Typical Hall experiment set up showing the direction of the current, I_x, and the magnetic field, B_z. V_y denotes the Hall voltage.

or yields the values for the fundamental constants used in quantum physics is truly amazing and hence calls for an intense scrutiny. The effect occurs when the density of the carriers, n, is such that they are encoded in the integers (that come as proportionality constants to the Hall resistivity in terms of h/e^2) as if the charges locked their separation at some particular values. The phenomenon remains resilient to changing the carrier density by a small amount; however, changing it by a large amount does destroy the effect.

The Hall resistivity (upper curve) in Fig. 1.3 becomes constant for certain ranges of the external magnetic fields, which are called plateaus. Further, the longitudinal resistivity (lower curve) in the same plot vanishes everywhere. Although it shows peaks wherever there is a jump in the Hall resistivity from one plateau to another. Later on, it was observed that ν may take values that are rational fractions, such as $\nu = \frac{1}{5}, \frac{2}{5}, \frac{3}{5}, \frac{3}{7}, \frac{4}{9}, \frac{5}{9}$. There are about 100 fractions (including the improper ones) that have been noted in experiments so far. The corresponding plot appears in Fig. 1.4.

1.2 General perspectives

The charge carriers being confined in 2D wells have a longer history. Since 1966 it has been known that the electrons accumulated at the surface of a silicon single

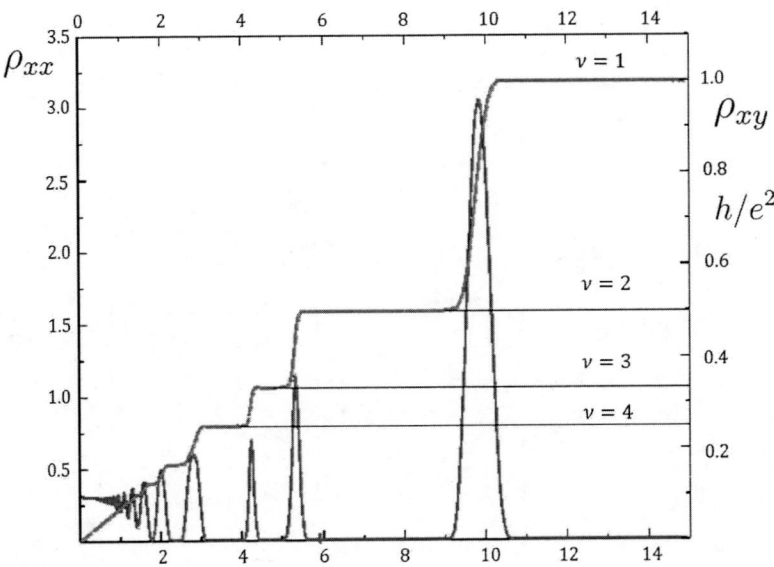

FIGURE 1.3. Schematic plot of integer quantum Hall effect (IQHE) as a function of the applied magnetic field. The plateaus in the upper curve denote quantization of Hall resistivity (ρ_{xy}), while the lower curve with spikes denote the magnetoresistivity (ρ_{xx}).

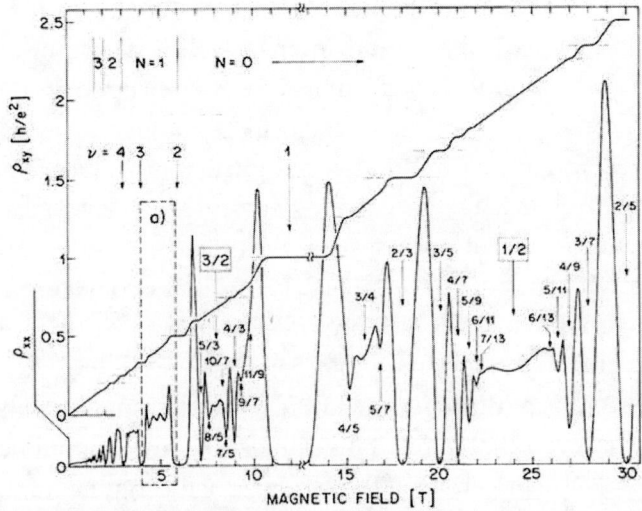

FIGURE 1.4. The plot shows fractional quantum Hall effect (FQHE). The plateaus are shown at fractional values in units of h/e^2. Taken from Ref.[62].

crystal induced by a positive gate voltage form a 2D electron gas (2DEG). The energy of the electrons that moves perpendicular to the surface is quantized (box quantization), and on top of it, the free motion of the electrons in 2D becomes quantized when a strong magnetic field is applied perpendicular to the plane

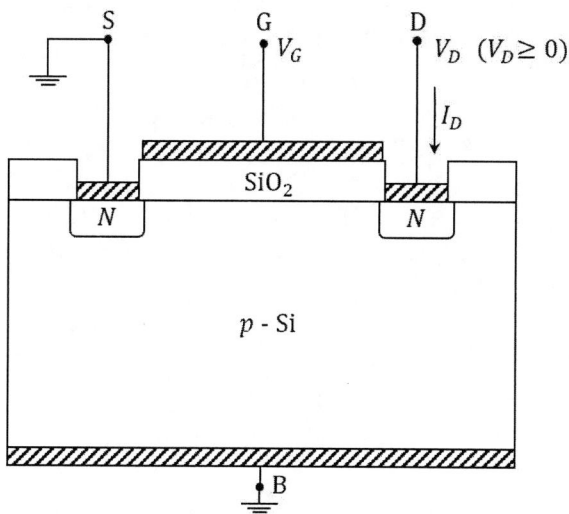

FIGURE 1.5. MOSFET structure showing base (B), gate (G), source (S) and drain (D). The substrate is a p-type Si. SiO_2 denotes the insulating oxide layer.

(Landau quantization). Thus, QHE has both the quantization phenomena built into it.

An important recent development in the study of semiconductors is the achievement of structures in which the electrons are restricted to move essentially in 2D. This immediately indicates that the carriers are prohibited to move along the direction transverse to the plane. Hence, the motion is quantized. Such 2D behaviour of the carriers can be found in metal-oxide-semiconductor (MOS) structures, quantum wells and superlattices. An excellent prototype is the metal-insulator-semiconductor (MIS) layered structure of which the insulator is usually an oxide, such as Al_2O_3 (thereby making it an MOS structure). In Fig. 1.5, we show a typical MOS device where the substrate is a doped p-type silicon (Si), which is grounded and is called a base (shown by B in Fig. 1.5). On the top, there is a metallic layer (shown by the hatched regime), followed by an insulating layer formed by SiO_2. The metallic layer is called the gate, denoted by G, which is biased by a voltage V_G. The source (grounded) and the drains are denoted by S and D, respectively, which in Fig. 1.5 consist of n-type materials. The gate voltage causes the carriers beneath the gate electrode to drift between the source and the drain. The layer of charge carriers below the oxide layer forms the 2DEG, which is central to our discussion. The energy dispersion in this case reads

$$E_n(k_x, k_y) = \frac{\hbar^2 k_x^2}{2m_{xx}^*} + \frac{\hbar^2 k_y^2}{2m_{yy}^*}, \tag{1.2.1}$$

where m^*_{xx} and m^*_{yy} are the components of the effective mass tensor defined by the inverse of the curvature of the band structure, namely

$$m^*_{\alpha\beta} = \hbar^2 \left(\frac{\partial^2 E(k)}{\partial k_\alpha \partial k_\beta} \right)^{-1}. \tag{1.2.2}$$

The physical properties of all systems are governed by their density of states (DOS), which plays a crucial role in deciding on the dependence of the temperature, density of carriers, etc. In 2D systems with a parabolic dispersion, as elaborated above in Eq. (1.2.1), the DOS is a constant and assumes a form

$$g(E) = g_{2D} = \frac{m^*}{\pi\hbar^2}. \tag{1.2.3}$$

The energy-independent DOS is very special to 2D and is in sharp contrast to three dimensions (3D), where it goes as $E^{1/2}$, and to one dimension (1D), where it goes as $E^{-1/2}$. In a general sense, and not restricted to the discussion on the Hall effect, the DOS enters while calculating the average quantities, such as the average energy or the average number of particles. For example, the average of a physical observable, O, of a fermionic system is computed using

$$\langle O \rangle = \int_0^\mu Of(E)g(E)dE, \tag{1.2.4}$$

where $f(E)$ is the Fermi distribution function given by

$$f(E) = \frac{1}{e^{\beta(E-\mu)} + 1},$$

with $\beta = \frac{1}{k_B T}$ and μ denoting the chemical potential. In general, this integral is quite challenging to compute analytically because of the Fermi distribution function ($f(E)$) present in the integrand.

Meanwhile, there is a wonderful simplification where $f(E)$ assumes a value unity at all temperatures at which the experiments are performed. Only at temperatures close to the Fermi temperature, T_F, defined via $\epsilon_F = k_B T_F$ (ϵ_F being the Fermi energy), $f(E)$ starts deviating from unity, and its exact form needs to be incorporated in the integral. However, T_F is usually of the order of tens of thousands of Kelvin for typical metals (such as Cu and Al), which is too high for them to appear in experimental situations. Moreover, the DOS only depends on energy and is independent of the temperature to a very good approximation. Thus, computation of Eq. (1.2.4) becomes independent or weakly dependent on temperature [63, 64].

1.3   Why is 2D important?

In the following discussion, we mention that there is something interesting about the transport properties of 2D systems. In the linear response regime Ohm's law is valid and indicates that

$$V_\alpha = R_{\alpha\beta} I_\beta, \tag{1.3.1A}$$

where $R_{\alpha\beta}$ denotes the resistivity tensor and α, β denote spatial variables x, y, etc. One can equivalently invert this equation to write $I_\alpha = G_{\alpha\beta} V_\beta$, where $G_{\alpha\beta}$ represents the conductivity tensor with $G = R^{-1}$. Equivalent relations in terms of the components of the electric field (**E**) and current density (**j**) read as

$$E_\alpha = \rho_{\alpha\beta} j_\beta \quad \text{and} \quad j_\alpha = \sigma_{\alpha\beta} E_\beta, \tag{1.3.1B}$$

where ρ and σ denote the resistivity and the conductivity tensors, respectively.

An interesting (and useful too) artefact of 2D physics is an accidental similarity that exists in decoding some of the key features of the transport properties. For example, the resistivity, ρ (or the conductivity, σ), is a quantity that is independent of the system geometry and hence is useful for a theoretical analysis. Whereas in experiments, one measures the resistance of a sample, R (or the conductance, G). For a sample in the shape of a hypercube of sides having length, L, the resistance and the resistivity are related by

$$R = \rho L^{2-d}, \tag{1.3.2}$$

where d denotes the dimensionality. Only for $d = 2$ is the resistance a scale invariant quantity. This puts the experimentalists and the theorists on the same page, as the geometry of the sample does not enter explicitly in the analysis of its transport properties.

1.4   Why are the conductivity and the resistivity tensors antisymmetric?

As we dig more into the details of the transport properties of 2DEG in the presence of a magnetic field, further useful information emerges. The off-diagonal elements of both the conductivity and the resistivity tensors are antisymmetric with regard to the direction of the applied field, **B**. Consider a planar sample with dimensions

$L_x \times L_y$. The conductivity tensor is of the form

$$\sigma = \begin{pmatrix} \sigma_{xx} & \sigma_{xy} \\ \sigma_{yx} & \sigma_{yy} \end{pmatrix}. \tag{1.4.1}$$

Let us try to understand the nature of the tensor, σ, in the presence of a magnetic field. A conductor in an external magnetic field obeys

$$j_\alpha = \sigma_{\alpha\beta} E_\beta. \tag{1.4.2}$$

Onsager's reciprocity principle does not hold in the presence of a magnetic field, **B**, which implies [65]

$$\sigma_{\alpha\beta}(\mathbf{B}) \neq \sigma_{\beta\alpha}(\mathbf{B}). \tag{1.4.3}$$

Instead, one has $\sigma_{\alpha\beta}(\mathbf{B}) = \sigma_{\beta\alpha}(-\mathbf{B})$ to make sure that the time reversal holds only if **B** changes sign. Let us write the conductivity tensor as a sum of a symmetric and an antisymmetric tensor (note that this is always possible for a rank 2 tensor). Thus,

$$\sigma_{\alpha\beta} = S_{\alpha\beta} + A_{\alpha\beta}, \tag{1.4.4}$$

where **S** and **A** are the symmetric and the antisymmetric tensors that obey the relations

$$S_{\alpha\beta}(\mathbf{B}) = S_{\beta\alpha}(-\mathbf{B}) = S_{\alpha\beta}(-\mathbf{B}) \tag{1.4.5}$$
$$A_{\alpha\beta}(\mathbf{B}) = A_{\beta\alpha}(-\mathbf{B}) = -A_{\alpha\beta}(-\mathbf{B})$$

such that the components of $S_{\alpha\beta}$ are even functions of **B**, while those of $A_{\alpha\beta}$ are odd functions of **B**. Putting in Eq. (1.4.2),

$$j_\alpha = S_{\alpha\beta} E_\beta + A_{\alpha\beta} E_\beta. \tag{1.4.6}$$

But owing to the antisymmetry,

$$A_{\alpha\beta} = \epsilon_{\gamma\alpha\beta} A_\gamma = -\epsilon_{\alpha\gamma\beta} A_\gamma. \tag{1.4.7}$$

Putting it in Eq. (1.4.6),

$$\begin{aligned} j_\alpha &= S_{\alpha\beta} E_\beta - \epsilon_{\alpha\beta\gamma} A_\beta E_\gamma \\ &= S_{\alpha\beta} E_\beta - (\mathbf{A} \times \mathbf{E})_\alpha \\ &= S_{\alpha\beta} E_\beta + (\mathbf{E} \times \mathbf{A})_\alpha. \end{aligned} \tag{1.4.8}$$

Assuming that we can expand $\sigma(\mathbf{B})$ in powers of $\mathbf{B}$,[1] such that the antisymmetric part contains odd powers of $\mathbf{B}$, we can write

$$A_\alpha = \eta_{\alpha\beta}B_\beta, \tag{1.4.9}$$

and $S_{\alpha\beta}(\mathbf{B})$ consists of even powers of $\mathbf{B}$,

$$S_\alpha = (\sigma_0)_{\alpha\beta} + \zeta_{\alpha\beta\gamma\delta}B_\gamma B_\delta. \tag{1.4.10}$$

The first term is the zero field conductivity tensor. Thus, putting things together up to terms linear in $\mathbf{B}$,

$$j_\alpha = S_{\alpha\beta}E_\beta + (\mathbf{E} \times \mathbf{A})_\alpha. \tag{1.4.11}$$

The second term denotes the Hall effect, which is linear in $\mathbf{B}$. This implies that the Hall current is perpendicular to the electric field, $\mathbf{E}$, and is proportional to $\mathbf{E}$ and $\mathbf{B}$. Thus, an antisymmetric tensor is relevant to the study of the Hall effect, which is why the conductivity and the resistivity tensors are antisymmetric. This is an important result that deviates from the corresponding scenario that arises in the absence of an external magnetic field.

1.5 Translationally invariant system: Classical limit of QHE

It is quite an irony that the extreme universal signature of the transport properties of a 2DEG is characterized by the flatness of the plateaus that not only occurs but survives even in the presence of disorder, impurity and imperfection. In the absence of the magnetic field, Anderson localization would have governed the transport signatures of non-interacting electrons, which indicates that in any dimension less than three, all eigenstates of a system are exponentially localized even for an infinitesimal disorder strength. Only in 3D is there a critical disorder at which a metal-insulation transition occurs. However, the scenario is strongly altered by the presence of the magnetic field, which yields, as we shall shortly see, a series of unique phase transitions from a perfect conductor to a perfect insulator. No other system demonstrates re-occurrence of the same phases over and over again as the magnetic field is gradually ramped up.

---

[1] It should be valid for weak magnetic fields and is not exactly true for the quantum Hall effect, but nevertheless it serves our purpose.

To begin with, we shall consider the case which is free from disorder, or equivalently a translationally (Lorentz) invariant system that possesses no preferred frame of reference. Thus, we can think of a reference frame that is moving with a velocity $-\mathbf{v}$ with respect to the lab frame, where the current density is given by $\mathbf{j} = -ne\mathbf{v}$ (n: areal electron density, $-e$: electronic charge). In this frame, the electric and the magnetic fields are given by[2]

$$\mathbf{E} = -\mathbf{v} \times \mathbf{B} \quad \text{and} \quad \mathbf{B} = B\hat{z}. \tag{1.5.1}$$

The above transformation ensures that an electric field must exist to balance the Lorentz force $-e\mathbf{v} \times \mathbf{B}$ in order to conduct without deflection. For the electric field, this yields

$$\mathbf{E} = \frac{1}{ne}\mathbf{J} \times \mathbf{B}. \tag{1.5.2}$$

This is equivalent to the tensor equation

$$E^{\mu} = \rho_{\mu\nu}j^{\nu}, \tag{1.5.3}$$

with the resistivity tensor given by

$$\rho_{\mu\nu} = \frac{B}{ne}\begin{pmatrix} 0 & 1 \\ -1 & 0 \end{pmatrix}. \tag{1.5.4}$$

Inverting the tensor equation, one can obtain

$$j^{\mu} = \sigma_{\mu\nu}E^{\nu}, \tag{1.5.5}$$

where the conductivity tensor $\sigma_{\mu\nu}$ is defined via

$$\sigma_{\mu\nu} = \frac{ne}{B}\begin{pmatrix} 0 & -1 \\ 1 & 0 \end{pmatrix}. \tag{1.5.6}$$

There is an interesting paradox that states that $\sigma_{xx} = \rho_{xx} = 0$ (see above), which is of course contradictory. However, we reserve this rather interesting topic for a discussion immediately afterwards. Here we wish to point out that we get the results for a classical Hall effect, that is, $\sigma_{xy} = \frac{ne}{B}$ (or $\rho = \frac{B}{ne}$). It is important to realize that the result is an artefact of Lorentz invariance, where the

---

[2]Remember that $\mathbf{E} = 0$ in the lab frame, though $\mathbf{B}$ remains unchanged.

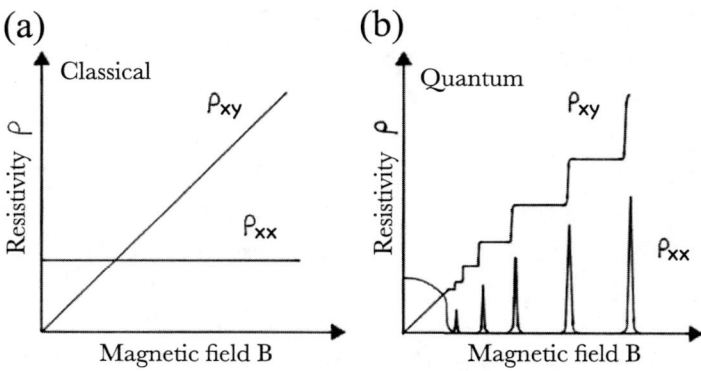

FIGURE 1.6. Schematic plot showing Hall and magnetoresistivities for both classical (a) and quantum (b) Hall effects.

characteristics of the sample or the 2DEG enter only through the carrier density, n, for a translationally invariant system. Thus, in the absence of defect, disorder and impurity, the Hall effect concerns with the carrier density of the sample and nothing else. The Hall resistivity depends linearly on the magnetic field. The QHE is more versatile (than merely depending on the density) where disorder (that usually jeopardizes the translational invariance) plays an indispensable role in the origin of the plateaus in conductivity (or resistivity) and their stability. We show a schematic plot in Fig. 1.6 to emphasize the difference between the classical and the quantum Hall effects. Both the Hall and the longitudinal resistivities are significantly different in these two cases.

Writing the equation of motion (EOM) for a charge particle of mass m, moving with a velocity $\mathbf{v}$ $(=\frac{\mathbf{p}}{m})$ in the presence of a longitudinal electric field $\mathbf{E}$ and a perpendicular magnetic field $\mathbf{B}$,

$$\frac{d\mathbf{p}}{dt} = -e\mathbf{E} - e\frac{\mathbf{p}}{m} \times \mathbf{B} - \frac{\mathbf{p}}{\tau}. \tag{1.5.7}$$

The last term is the resistive force arising from electron-impurity scattering, with τ being the relaxation time. Using the current density, $\mathbf{j} = -ne\mathbf{p}/m$, and the cyclotron frequency, $\omega_B = eB/m$, at the steady state $(\frac{d\mathbf{p}}{dt} = 0)$,

$$\frac{ne^2\tau}{m}\mathbf{E} + \omega_c\tau\mathbf{j} \times \hat{\mathbf{z}} + \mathbf{j} = 0. \tag{1.5.8}$$

Assuming the motion of carriers along the x-direction, that is, $\mathbf{j} = j\hat{\mathbf{x}}$, casting it in the form $\mathbf{j} = \sigma\mathbf{E}$, σ assumes the form

$$\sigma = \frac{ne^2\tau/m}{1+\omega_c^2\tau^2} \begin{pmatrix} 1 & -\omega_c\tau \\ \omega_c\tau & 1 \end{pmatrix}. \tag{1.5.9}$$

This yields the Drude conductivity, which can be written as

$$\sigma_{xx} = \frac{\sigma_0}{1 + \omega_c^2 \tau^2},$$ (1.5.10)

where $\sigma_0 = ne^2\tau/m$. In the absence of any scattering by the impurities, the relaxation time, τ, is infinitely large, which yields $\sigma_{xx} \to 0$. This induces Ohm's law to assume the form (now writing in terms of the resistivity)

$$\mathbf{E} = \begin{pmatrix} E_x \\ E_y \end{pmatrix} = \begin{pmatrix} 0 & \rho_{xy} \\ -\rho_{xy} & 0 \end{pmatrix} \begin{pmatrix} j_x \\ j_y \end{pmatrix}$$
$$= \begin{pmatrix} \rho_{xy} j_y \\ -\rho_{xy} j_x \end{pmatrix}.$$ (1.5.11)

Hence, the electric field $\mathbf{E}$ is perpendicular to the current density $\mathbf{j}$, which indicates that $\mathbf{j} \cdot \mathbf{E} = 0$. The physical significance of $\mathbf{j} \cdot \mathbf{E}$ is the work done that accelerates the charges, which being zero in this case implies that a steady current flows in the sample without requiring any work, and hence causes no dissipation. Thus, $\sigma_{xx} = 0$ implies that no current flows in the longitudinal direction, which is actually a signature of a perfectly insulating state. Since the components of the resistivity and the conductivity tensors are related by

$$\sigma_{xx} = \frac{\rho_{xx}}{\rho_{xx}^2 + \rho_{xy}^2} \quad ; \quad \sigma_{xy} = \frac{-\rho_{xy}}{\rho_{xx}^2 + \rho_{xy}^2}.$$ (1.5.12)

Let us examine the possible scenarios:

(i) If $\rho_{xy} = 0$, one gets $\sigma_{xx} = \frac{1}{\rho_{xx}}$ and $\sigma_{xy} = 0$, which is a familiar scenario.

(ii) If $\rho_{xy} \neq 0$, σ_{xx} and σ_{xy} both exist.

(iii) Now consider $\rho_{xx} = 0$, $\sigma_{xx} = 0$, if $\rho_{xy} \neq 0$. This is truly interesting, since if $\rho_{xx} = 0$, it implies a perfect conductor, and at the same time $\sigma_{xx} = 0$ implies a perfect insulator. This is surprising but truly occurs in the presence of an external magnetic field. This is reflected in the plots presented in Figs. 1.3 and 1.4. We shall return for a more thorough discussion later.

1.6   Charge particles in a magnetic field: Landau levels

Let us examine the fate of the electrons confined in a 2D plane in the presence of a magnetic field. Consider non-interacting spinless electrons in an external field, $\mathbf{B}$, and write down the Schrödinger equation to solve for the eigenvalues and eigenfunctions. The canonical momentum can now be written as $\mathbf{p} \to \mathbf{p} - q\mathbf{A} = \mathbf{p} + e\mathbf{A}$ [66], where $q = -e$ is the electronic charge and $\mathbf{A}$ is the vector potential corresponding to the field, $\mathbf{B}$. $\mathbf{B}$ and $\mathbf{A}$ are related via $\nabla \times \mathbf{A} = \mathbf{B}$. The time-independent Schrödinger equation becomes

$$\frac{1}{2m}(\mathbf{p} + e\mathbf{A})^2 \psi(\mathbf{r}) = E\psi(\mathbf{r}). \tag{1.6.1}$$

In order to solve this, we need to fix a gauge or a choice of the vector potential. Note that the choice of the gauge will not alter the solution of the equation, which in other words can be stated as the Schrödinger equation being gauge invariant. However, a particular choice is essential for us to go ahead.

Corresponding to a magnetic field, $\mathbf{B}$, in the z-direction such that $\mathbf{B} = B\hat{z}$ (just as in the case of a 2DEG subjected to a perpendicular magnetic field), the vector potential can be chosen as

$$A_x = -By, \quad A_y = A_z = 0. \tag{1.6.2}$$

This is known as the Landau gauge. It also allows us to assume $A_y = Bx$ and $A_x = A_z = 0$.[3] In the Landau gauge, the Schrödinger equation becomes

$$\left[\frac{1}{2m}(p_x - eBy)^2 + \frac{p_y^2}{2m} + \frac{p_z^2}{2m} \right] \psi(\mathbf{r}) = \epsilon \psi(\mathbf{r}). \tag{1.6.3}$$

Clearly, in the z-direction, the particle behaves like a free particle with energy $\epsilon_z = \frac{p_z^2}{2m}$, and with the eigenfunction same as that of a particle in a box in the z-direction. Thus, in the x–y plane, the above equation becomes

$$\left[\frac{1}{2m}(p_x - eBy)^2 + \frac{p_y^2}{2m} \right] g(x,y) = \epsilon g(x,y) \tag{1.6.4}$$

$$\mathcal{H}(x,y) g(x,y) = \epsilon g(x,y),$$

where for the 2D case, $\psi(\mathbf{r})$ becomes $g(x,y)$ and $E = \epsilon + \epsilon_z$. It is easy to see that p_x commutes with $\mathcal{H}(x,y)$, that is,

$$[\mathcal{H}(x,y), p_x] = 0. \tag{1.6.5}$$

---

[3] A combination of the two yields a 'symmetric' gauge, which we shall introduce and employ later.

Hence, p_x is a constant of motion. Thus, for a p_x given by $p_x = \frac{2\pi\hbar}{L_x}n_x$, one can write Eq. (1.6.5) as

$$\left[\frac{p_y^2}{2m} + \frac{1}{2}m\left(\frac{eB}{m}\right)^2(y-y_0)^2\right]f(y) = \epsilon f(y), \tag{1.6.6}$$

where $y_0 = \frac{p_x}{eB} = k_x l_B^2 = k l_B^2$ (say) and $f(y)$ is only a function of y. y_0 has the dimension of length, and hence l_B is denoted as the magnetic length, which is an important quantity in all subsequent discussions. Moreover, $f(y)$ is the eigenfunction corresponding to the 1D Hamiltonian written above.

Interestingly, the left-hand side of Eq. (1.6.6) denotes the Hamiltonian for a simple harmonic oscillator (SHO) that oscillates in the y-direction about a mean position y_0 with a frequency $\omega_c = \frac{eB}{m}$. ω_c is known as the cyclotron frequency. Taking results from the SHO problem in quantum mechanics, the energy eigenvalues can be found as

$$\epsilon_n = (n+\frac{1}{2})\hbar\omega_c = (n+\frac{1}{2})\hbar\frac{eB}{m}. \tag{1.6.7}$$

The energy levels in Eq. (1.6.7) are referred to as the Landau levels. The levels are schematically shown in Fig. 1.7. They are equidistant for a given value of the magnetic field; however, their separation increases with the increase in B. Each of these levels has a large degeneracy (see below).

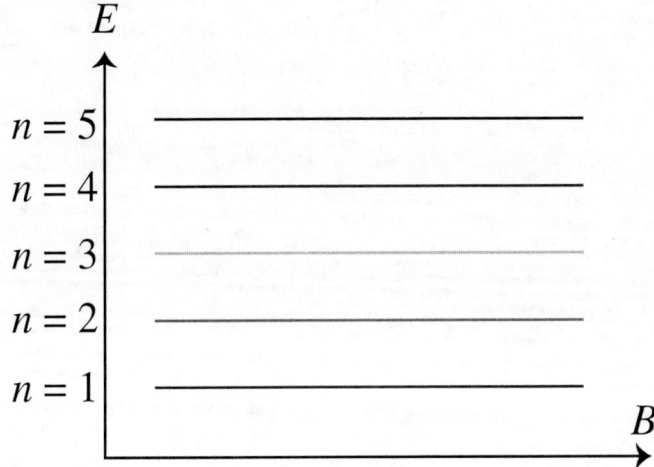

FIGURE 1.7.   Schematic plot showing the Landau levels. Each of these levels has a large degeneracy.

Further, the eigenfunction, $g(x, y)$, in Eq. (1.6.5) corresponding to an oscillatory motion (as an SHO) in the y-direction and a free motion (like a particle in the absence of any potential) in the x-direction assumes the form

$$g(x, y) = \frac{1}{L_x} e^{ik_x x} A_n e^{\frac{-eB(y-y_0)^2}{\hbar}} H_n\left(\frac{eB(y - y_0)}{\hbar}\right),\qquad (1.6.8)$$

where $H_n(\xi)$ with $\xi = \frac{eB(y-y_0)}{\hbar}$ denote the Hermite polynomials that are familiar in the context of SHO, and A_n denote the normalization constants. Thus, the trajectory of the particle is similar to that of an SHO centred about a certain value of y (instead of the origin), namely y_0 and freely propagating along the x-direction. y_0 is controlled by the strength of the magnetic field B and is inversely proportional to it.

1.7 Degeneracy of the Landau levels

The Landau levels given by Eq. (1.6.7) are hugely degenerate. Since p_x is a constant of motion, the energy is independent of p_x. Thus, all possible values of the quantum numbers corresponding to the motion in the x-direction, namely n_x that are defined by $k_x = \frac{2\pi}{L_x} n_x$ (L_x denotes the length of the sample in the x-direction and $n_x = 0, 1, 2 \ldots$) will make the levels degenerate. The degree of degeneracy is only limited by the length of the sample in the y-direction, namely L_y. Since the magnetic length $y_0 = k l_B^2 = k\left(\frac{\hbar}{eB}\right)$ about which the simple harmonic motion occurs should not exceed the length L_y of the sample, the maximum degeneracy of the Landau levels can be found by substituting $y_0 = L_y = \left(\frac{\hbar}{eBL_x}\right) n_x$ and hence using the maximum possible value of n_x, that is,

$$(n_x)_{max} = g = \frac{eBL_xL_y}{h} = \frac{eB\mathcal{A}}{h},\qquad (1.7.1)$$

where $\mathcal{A} = L_xL_y$ is the area of the sample in the x–y plane. This yields the degeneracy, g, to be identified as the flux, ϕ ($=B\mathcal{A}$), threading the planar sample via $g = \Phi/\Phi_0$, where $\Phi_0 = h/e$.[4]

A few comments are in order:

(i) The degeneracy, g, is independent of the effective mass of the carriers and hence independent of the material.

---

[4]The value of $\phi_0 \simeq 4.13 \times 10^{-15}$ Wb.

(ii) The degeneracy is proportional to the area of the sample and the value of the magnetic field. Thus, the degeneracy can be controlled by the applied magnetic field.

To remind ourselves, we have solved for the properties of a single electron confined in a plane in the presence of a perpendicular magnetic field. The energies are called the Landau levels, and these levels are highly degenerate. The scenario is a prototype of what happens in a Hall effect experiment where the external magnetic field is varied and the resistivities (both Hall and the longitudinal) are measured. The Hall resistivity shows plateaus in multiples of h/ne^2 whenever the filling fraction, ν, of the Landau levels (defined below) is close to an integer n. Consider n_0 to be the density of charge carriers of the sample; then ν is defined using $\nu = \frac{n_0}{g/A} = \frac{n_0 h}{eB}$. ν denotes the filling fraction, which is also defined as

$$\nu = \frac{\text{number of electrons}}{\text{number of flux quanta}}.$$

As and when ν assumes a value near an integer, n (or a rational fraction as in the case of FQHE), as the magnetic field is tuned, one observes a plateau in the Hall resistivity, ρ_{xy}. Something else happens at the same time that is equally interesting. The longitudinal resistivity, ρ_{xx}, drops to zero whenever ρ_{xy} acquires a plateau. The vanishing of ρ_{xx} makes the system dissipationless. However, the diagonal conductivity (σ_{xx}) also vanishes, which makes the system insulating. The paradox is explained in detail elsewhere.

1.8 Conductivity of the Landau levels

Here we discuss some of the key properties of the Landau levels. Let us now calculate the current carried by the Landau levels. The expression for the current can be found using $\langle \mathbf{J} \rangle = -e \langle \mathbf{v} \rangle$, where the expectation value has to be computed within the Landau states.

$$\langle \mathbf{J} \rangle = -e \langle \psi_k | \mathbf{v} | \psi_k \rangle = -\frac{e}{m} \langle \psi_k | (\mathbf{p} + e\mathbf{A}) | \psi_k \rangle. \qquad (1.8.1)$$

The longitudinal current in the x-direction carried by the Landau levels is obtained via

$$\langle J_x \rangle = -\frac{e}{m l_B \sqrt{\pi}} \int dy \, e^{-\frac{1}{l_B^2}(y-y_0)^2} (\hbar k - eBy) = 0. \qquad (1.8.2)$$

The integrand has an even function (the first one) and an odd function (the second one). Thus, the integral vanishes. We can get the average velocity, $\langle v \rangle = \frac{1}{\hbar} \frac{\partial \epsilon_k}{\partial k} = 0$, as ϵ does not depend upon k. Thus, the Landau wavefunctions carry no current by themselves. They only carry current in the presence of an electric field in the x-direction as we show below.

1.9 Spin and the electric field

So far we have been talking about spinless fermions. It is in general a worthwhile exercise to include the spin of the electrons and explore if there is any significant development to the quantization phenomena discussed above. The spin degrees of freedom placed in an external magnetic field introduces a Zeeman energy scale owing to a coupling between the spin and the magnetic field. The Zeeman term is written as $\Delta_Z = g\mu_B B$, where $\mu_B = e\hbar/2m$ is the Bohr magneton and $g = 2$. Now the splitting between the Landau levels originating from the orbital effect ($\mathbf{p} \to \mathbf{p} - e\mathbf{A}$) is $\Delta = \hbar\omega_B = \frac{e\hbar B}{m} = \Delta_B (say)$. Δ_B is the so-called cyclotron energy. But for electrons, this precisely coincides with the Zeeman splitting $\Delta = g\mu_B B$ between the $\uparrow$ and the $\downarrow$-spins. Thus, it looks like the spin-$\uparrow$ particles with the Landau level index n and the spin-$\downarrow$ particles with the next higher Landau level index (that is, $n + 1$) have the same energies. However, in real materials this does not occur. For example in GaAs, the Zeeman energy is typically about 70 times smaller than the cyclotron energy.

Now consider an external electric field, E, applied in the x-direction. This creates an electric potential of the form $\phi = -Ex$. Thus, the resulting Hamiltonian including external electric and magnetic fields can be written as

$$\mathcal{H} = \frac{1}{2m}\left[p_x^2 + (p_y + eBx)^2\right] + eEx. \tag{1.9.1}$$

It is important to note that we have considered a different choice of the gauge here, namely $A_y = Bx$. Instead of a rigorous derivation, we can complete the square in the expression for energy of the particles

$$E_{n,k} = \hbar\omega_B\left(n + \frac{1}{2}\right) - eE\left(kl_B^2 + \frac{eE}{m\omega_B^2}\right) + \frac{1}{2}m\frac{E^2}{B^2}. \tag{1.9.2}$$

The middle term yields a linear in k dependence, and the spectrum is shown in Fig. 1.8. This is interesting because the degeneracy of the Landau level has now been lifted. The energy in each level now depends linearly on k as shown, which was earlier independent of k. The eigensolution is simply that of a harmonic oscillator shifted from the origin and displaced along the x-axis by an amount

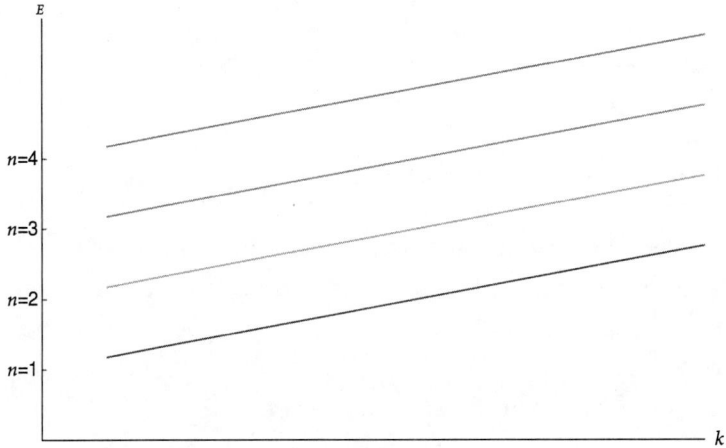

FIGURE 1.8. Schematic plot showing the Landau levels ($n = 1, 2, 3, 4$) in the presence of an electric field. The levels are tilted because of the electric field.

mE/eB^2 and is written as

$$\psi(x,y) = \psi_{n,k}\left(x + \frac{mE}{eB^2}, y\right).$$ (1.9.3)

Among the other properties, the group velocity is given by

$$v_y = \frac{1}{\hbar}\frac{\partial E_{n,k}}{\partial k} = \frac{e}{\hbar}El_B^2.$$ (1.9.4)

Putting $l_B = \sqrt{\frac{\hbar}{eB}}$ (as said earlier, l_B is an important length scale of the problem which we shall see throughout the discussion),

$$v_y = \left(\frac{eE}{\hbar}\right)\cdot\left(\frac{\hbar}{eB}\right) = \frac{E}{B}.$$ (1.9.5)

Thus, the energy has three terms in Eq. (1.9.2):

(i) the first one is that of a harmonic oscillator,
(ii) the second one is the potential energy of a wave packet localized at $x = (-kl_B^2 - eE/m\omega_B^2)$, and
(iii) the last one denotes the kinetic energy of the particle, namely $\frac{1}{2}mv_y^2$.

1.10   Laughlin's argument: Corbino ring

Laughlin intuitively considered the phenomenon of QHE as a quantum pump. Consider a ring where the vacant region admits a magnetic field and hence a flux

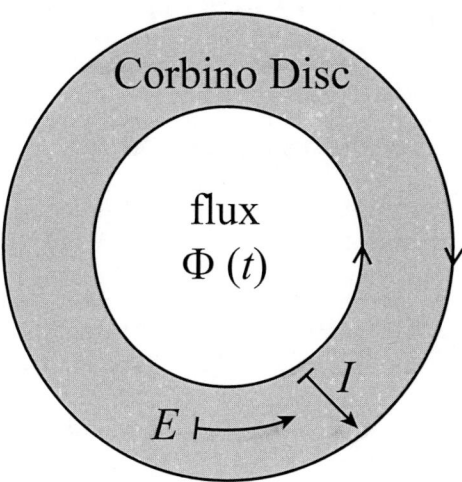

FIGURE 1.9. Schematic plot showing Corbino ring.

ϕ (see Fig. 1.9). For the argument to be valid, the geometry of the ring is important. Here in addition to the background magnetic field **B** that threads the sample, we can thread an additional flux Φ through the centre of the ring. This Φ can affect the quantum state of the electrons. In addition, the temperature is low such that the thermal effects can be neglected.

Let us first see what this flux Φ has got to do with the Hall conductivity. Suppose we slowly increase Φ from 0 to $\Phi_0 (= \frac{h}{e})$, that is, within a time $t_0 \gg \frac{1}{\omega_B}$. This induces an emf around the annular region $\varepsilon = \frac{\partial \Phi}{\partial t} = \frac{-\Phi_0}{t_0}$. The purpose of this emf is to transport 'n' electrons from the inner circumference to the outer circumference. This would result in a current in the radial direction, $I_r = -ne/t_0$. Thus, the Hall resistivity is

$$\rho_{xy} = \frac{\varepsilon}{I_r} = -\frac{\Phi_0}{t_0} \cdot \frac{t_0}{(-ne)} = \frac{h}{e^2} \cdot \frac{1}{n}. \tag{1.10.1}$$

The same arguments hold equally for the IQHE and FQHE; in the former n is an integer, while n is a fraction for the latter. In FQHE, the interpretation is as follows: as we increase the flux from Φ to Φ_0, a charge of magnitude e/m is transported from the inner circumference to the outer one when the flux is increased by Φ_0 units. The resultant Hall conductivity (or equivalently the resistivity) becomes

$$\sigma_{xy} = \frac{e^2}{h} \cdot \frac{1}{m}. \tag{1.10.2}$$

Thus, a whole electron is transferred only when the flux is increased by $m\Phi_0$ units.

1.11 Edge modes and conductivity of the single Landau level

When a particle is restricted to move only in one direction, the motion is said to be chiral where backscattering is prohibited. Thus, the particles propagate in one direction at one edge of the sample and move in the other direction at the other end of the sample. Let us understand how the edge modes appear.

An edge can be modelled by a potential $V(x)$ in the y-direction which rises steeply as shown in Fig. 1.10. Let us continue working on the Landau gauge, such that the Hamiltonian is given by

$$\mathcal{H} = \frac{1}{2m}[p_x^2 + (p_y + eBx)^2] + V(x). \tag{1.11.1}$$

In the absence of the potential $V(x)$, the (lowest) wavefunction is a Gaussian of width l_B $\left(= \sqrt{\frac{\hbar}{eB}}\right)$. If we assume that $V(x)$ is smooth over a distance l_B, and hence assume the centre of each Gaussian to be localized at $x = x_0$, we can Taylor expand the potential $V(x)$ about x_0 in the following fashion:

$$V(x) = V(x_0) + \frac{\partial V}{\partial x}(x - x_0) + \cdots. \tag{1.11.2}$$

Dropping terms after the second one and assuming the constant term (first term) to be zero, the second term looks like the potential due to the electric field. So the particle acquires a drift velocity in the y-direction,

$$v_y = -\frac{1}{eB}\frac{\partial V}{\partial x}. \tag{1.11.3}$$

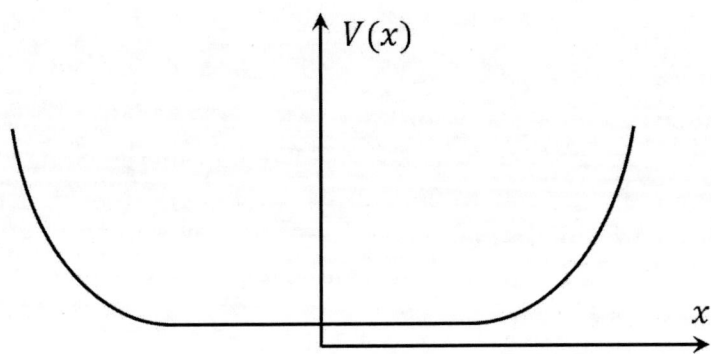

FIGURE 1.10. Schematic plot showing the potential seen by the charge due to edge of the quantum Hall sample.

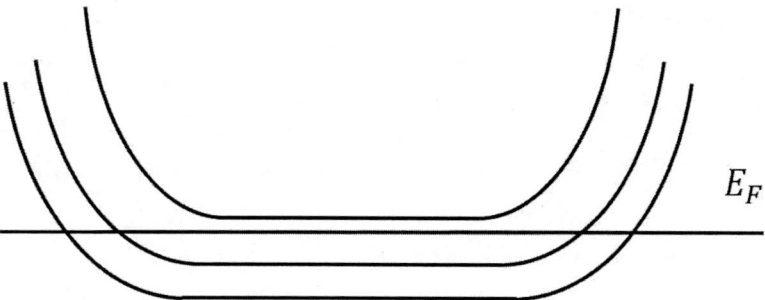

FIGURE 1.11.   Schematic plot showing the appearance of edge modes.

Each wavefunction labelled by a momentum k is located at different x-positions, namely $x = -kl_B^2$, and thus has a different drift velocity. Additionally, the slopes of the two edges drawn above have different slopes, so $\frac{\partial V}{\partial x}$ has different signs at the edges. Thus, v_y at the left edge has a different sign with respect to the right edge. Further, because of the drift in the y-direction, there will be a current I_y, which is known as the Hall current and is calculated as

$$
\begin{aligned}
I_y &= -e \int \frac{dk}{2\pi} v_y(k) \\
&= \frac{e}{2\pi l_B^2} \int dx \frac{1}{eB} \frac{dV}{dx}; \quad \text{using } l_B^2 = \frac{\hbar}{eB} \\
&= \frac{e^2}{2\pi\hbar} V_H.
\end{aligned}
\tag{1.11.4}
$$

V_H is the Hall voltage. Now, $\sigma_{xy} = \frac{I_y}{V_H} = \frac{e^2}{2\pi\hbar} = \frac{e^2}{h}$, which is indeed the expected conductivity for a single Landau level.

The schematic diagram in Fig. 1.11 shows that the current is entirely carried by the edge states, since the bulk Landau level is absolutely flat (having no k dependence) and hence does not carry any current. The argument is also elegant as it does not depend upon the form of the potential $V(x)$.

Everything that we have discussed so far holds for a single Landau level; however, the argument is equally valid for a large number of Landau levels as long as the Fermi energy lies in between the filled and the unfilled Landau levels. Also the chiral edge modes are robust to any impurity or disorder as there is no phase space available for scattering. If a left-moving electron is to scatter over to a right-moving electron, it has to cross the entire sample edge, which is not allowed as the probability of scattering would be infinitesimally small owing to the macroscopic physical dimension of the sample. Thus, the Hall plateaus are robust to disorder, defects and impurities.

A supremely important issue stands out: how do the plateaus exist in the first place? To see this, let us fix the electron density, n. Then we shall only have filled Landau levels when the magnetic field is exactly $B = \frac{n}{\nu}\Phi_0$ (with $\Phi_0 = \frac{h}{e}$) for some integer ν. But what happens when $B \neq \frac{n}{\nu}\Phi_0$, that is, when the Landau levels are partially filled? Also on top of that there is also a (small) electric field. In the partially filled last Landau level, the longitudinal conductivity will be non-zero, while the Hall conductivity will not be quantized. So how do the plateaus appear and the longitudinal conductivities vanish? The disorder comes to the rescue. It gives a finite width to the Landau levels. So even if the $B \neq \frac{n}{\nu}\Phi_0$, there is a finite regime over which the Hall conductivity, σ_{xy}, remains constant and hence the plateaus form so that the Hall resistivity, ρ_{xy}, freezes at that value, till the magnetic field is increased further.

1.12   Incompressibility of the quantum Hall states

The key feature of the QH states is that the states are incompressible. The compressibility, κ, is defined by

$$\kappa = -\frac{1}{A}\frac{\partial A}{\partial P}\Big|_N, \tag{1.12.1}$$

where A, P and N are the area, pressure and the number of particles, respectively. The system is said to be incompressible when $\kappa = 0$, that is, when the area is insensitive to the pressure applied. The pressure is defined as the change of energy as a result of change of area, that is,

$$P = -\frac{\partial E}{\partial A}. \tag{1.12.2}$$

Thus, the inverse of the compressibility is defined as

$$\kappa^{-1} = -A\frac{\partial P}{\partial A}\Big|_N = A\frac{\partial^2 E}{\partial A^2}\Big|_N. \tag{1.12.3}$$

Since the energy is an extensive quantity, that is, it depends on the number of particles and hence it can be written as

$$E = N\epsilon(n), \tag{1.12.4}$$

where ϵ is the energy per particle and n denotes the particle density. The total number of particles is given by $N = An$. Hence,

$$\kappa^{-1} = \frac{1}{n}\frac{d}{d(\frac{1}{n})}\frac{d\epsilon(n)}{d(\frac{1}{n})}$$
$$= n^2\left(2\frac{d\epsilon(n)}{dn} + n\frac{d^2\epsilon(n)}{dn^2}\right)$$
$$= n^2\frac{d^2(n\epsilon)}{dn^2}. \tag{1.12.5}$$

Also, the chemical potential is given by

$$\mu = \left.\frac{\partial E}{\partial N}\right|_V = \frac{d(n\epsilon)}{dn}. \tag{1.12.6}$$

Thus, comparing Eqs. (1.12.5) and (1.12.6),

$$\kappa^{-1} = n^2\frac{d\mu}{dn}. \tag{1.12.7}$$

The system is incompressible ($\kappa = 0$) when the chemical potential μ increases discontinuously as a function of density, that is, $\frac{\partial n}{\partial \mu} = 0$.

1.13 Derivation of the Hall resistance

Before we go on to derive the conductivity (or the resistivity) using the Kubo formula, let us present a simpler derivation. Consider a length, l, of the sample. The electric current carried by each charge in this length is given by $-ev/l$, where v denotes the group velocity and e is the magnitude of the electronic charge. Now the total number of electrons between momentum range p and $p + dp$ can be found by multiplying the current carried per unit charge ($= -ev/l$) and $(l/h)dp$, where h is Planck's constant. Thus, the elemental current is given by

$$dI = -(ev/l) \times \left(\frac{l}{h}\right)dp = -\frac{ev}{h}dp.$$

Thus, the arbitrary length l gets cancelled. Replacing the velocity by $v = \frac{dE}{dp}$, the elemental current becomes

$$dI = -\frac{e}{h}\frac{dE}{dp}dp. \tag{1.13.1}$$

Hence, the total current can be obtained by integrating Eq. (1.13.1) from a certain p_1 to p_2, where p_1 and p_2 denote momenta of the charges corresponding to the top

(TE) and bottom (BE) edges, respectively, of a typical Hall sample (see Fig. 1.2). Thus, the total current is given by

$$I = \int_{p_1}^{p_2} \left(-\frac{e}{h}\right) \left(\frac{dE}{dp}\right) dp \tag{1.13.2}$$
$$= -\frac{e}{h}[V(BE) - V(TE)],$$

where V denotes the potential energies at the two edges. Thus, the current is driven by the potential energy difference between the BE and the TE, which of course arises out of the potential difference that exists between them.[5] Thus,

$$I = -\frac{e}{h}\left[(-eV_2) - (eV_1)\right], \tag{1.13.3}$$

where V_2 and V_1 denote the voltages at the BE and TE, respectively. This yields a familiar form for the current, that is,

$$I = \frac{e^2}{h}(V_2 - V_1). \tag{1.13.4}$$

This is the Hall current, which is independent of the dimension of the sample. The resistance, or more precisely the Hall resistance, is given by

$$R_H = \frac{V_2 - V_1}{I} = \frac{h}{e^2},$$

which is precisely the Hall resistivity for the filling fraction $v = 1$. For an arbitrary filling fraction v, the number of electrons will be v times what it is for $v = 1$, where the Hall resistance can be written as

$$R_H = \frac{h}{ve^2}. \tag{1.13.5}$$

1.14   Kubo formula and the Hall conductivity

The important question at this stage is: what protects the plateaus in the Hall conductivity (or the resistivity)? Why are they so flat and robust at the integer values (or at certain rational fractions)? Remember, the system does not have either translational invariance (because of the presence of disorder) or time

---

[5]The top and the bottom edges are perpendicular to the direction of the applied electric field.

reversal invariance (because of the presence of a magnetic field). Thus, two known symmetries are lost and still the plateaus persist.[6] We shall now show that the Hall conductivity assumes $\sigma_{xy} = \frac{ve^2}{h}$ (or equivalently $\rho_{xy} = \frac{h}{ve^2}$) under these conditions.

To compute the conductivity, we resort to the Kubo formula, which arises from a more generalized concept, known as the *linear response theory*. We shall derive the Kubo formula under a few conditions for the sake of simplicity. They are as follows:

(i) Before the fields are applied, at $t = -\infty$, the system is in a non-interacting many particle state that obeys $\mathcal{H}_0 |\psi_m\rangle = E_m |\psi_m\rangle$, where $\{E_m, \psi_m\}$ are the eigensolutions of $\mathcal{H}_0$.

(ii) Even if we actually apply a constant electric field, **E**, it is helpful to consider an alternating field with frequency ω of the form $\mathbf{E(t)} = \mathbf{E}e^{-i\omega t}$ and at the end of the calculation take the zero frequency limit, that is, $\omega \to 0$.

(iii) Consider a gauge in which the transverse components of the vector potential are zero, that is, $A_t = 0$. In other words, $\mathbf{E} = -\frac{\partial \mathbf{A}}{\partial t}$ with no $\nabla\phi$ term, ϕ being the scalar potential. Equivalently, one can assume that $\phi = $ constant.

Now let us write the full Hamiltonian as

$$\mathcal{H} = \mathcal{H}_0 + \mathcal{H}', \tag{1.14.1}$$

where $\mathcal{H}_0$ is the non-interacting Hamiltonian whose exact solutions are known (as described earlier) and $\mathcal{H}'$ is the interaction term due to the coupling of the electrons to the external field. Thus, $\mathcal{H}'$ involves the current due to the motion of the electrons coupling with the vector potential arising due to the presence of the magnetic field. Thus,

$$\mathcal{H}' = -\mathbf{J} \cdot \mathbf{A}, \tag{1.14.2}$$

where **J** and **A** are the electric current density and the vector potential, respectively. **J** is related to the mechanical momentum $\mathbf{p} + e\mathbf{A}$. Using $\mathbf{E} = -\frac{\partial \mathbf{A}}{\partial t}$, one can write

$$\mathbf{A} = \frac{\mathbf{E}}{i\omega}e^{-i\omega t}. \tag{1.14.3}$$

The aim is to compute the expectation value of the current density and find out how it depends on the applied electric field such that the proportionality constant yields the conductivity. In particular, we are interested in computing the Hall conductivity, σ_{xy}.

---

[6]In fact, naively it seems ironical that a broken time reversal symmetry is solely responsible for the quantization of the Hall plateaus.

Here we shall consider the interaction picture where the time evolution of an arbitrary operator, $\hat{O}$, is written as

$$\hat{O}(t) = e^{i\mathcal{H}_0 t/\hbar}\hat{O}(0)e^{-i\mathcal{H}_0 t/\hbar}. \tag{1.14.4}$$

Here $\hat{O}$ can be any operator, such as $\mathbf{J}$ or $\mathcal{H}'$. Further, the eigenstates in the interaction picture evolve with time according to

$$|\psi(t)\rangle = U(t, t_0)|\psi(t_0)\rangle, \tag{1.14.5}$$

where t_0 refers to an earlier time when the interaction is switched on and t denotes a later time. The time evolution operator, $U(t, t_0)$, is a unitary operator having a form

$$U(t, t_0) = T\exp\left(-\frac{i}{\hbar}\int_{t_0}^{t}\mathcal{H}'(t')dt'\right). \tag{1.14.6}$$

T denotes time ordering in the above equation [68]. If the interval $[t : t_0]$ is split into several time steps, T keeps the earliest time to the right. Now let us consider that as $t_0 \to -\infty$, that is, before the perturbation is switched on, the system was in the ground state $|\psi_0(t)\rangle$. Hence, the time evolution operator can be written as

$$U(t, t_0 = -\infty) = T\exp\left(-\frac{i}{\hbar}\int_{-\infty}^{t}\mathcal{H}'(t')dt'\right) = U(t) \text{ (say)}. \tag{1.14.7}$$

The ground state expectation value of the current operator is given by

$$\begin{aligned}\langle \mathbf{J}(t)\rangle &= \langle \psi_0(t)|\mathbf{J}(t)|\psi_0(t)\rangle \\ &= \langle \psi_0|U^{-1}(t)\mathbf{J}(t)U(t)|\psi_0\rangle \\ &= \langle \psi_0|\left[Te^{\frac{i}{\hbar}\int_{-\infty}^{t}\mathcal{H}'(t')dt'}\mathbf{J}(t)Te^{-\frac{i}{\hbar}\int_{-\infty}^{t}\mathcal{H}'(t')dt'}\right]|\psi_0\rangle. \end{aligned} \tag{1.14.8}$$

An expansion of the exponentials (assuming the interaction term to be weak) and retaining terms up to first order in $\mathcal{H}'$ yields

$$\langle \mathbf{J}(t)\rangle \approx \langle \psi_0|\left[\mathbf{J}(t) + \frac{i}{\hbar}\int_{-\infty}^{t}dt'[\mathcal{H}'(t'), \mathbf{J}(t)]\right]|\psi_0\rangle. \tag{1.14.9}$$

The second term inside the bracket of the right-hand side (RHS) involves a commutator of $\mathcal{H}'$ and $\mathbf{J}$. It is to be kept in mind that the commutator does not vanish at two arbitrary times t and t'. Using $\mathbf{A}(t) = \frac{\mathbf{E}}{i\omega}e^{-i\omega t}$ and $\mathcal{H}'(t) = -\mathbf{J} \cdot \mathbf{A}$, the $\langle \mathbf{J}(t) \rangle$ takes the form

$$\langle \mathbf{J}(t) \rangle \approx \langle \psi_0 | \left[\mathbf{J}(t) + \frac{i}{\hbar} \int_{-\infty}^{t} dt' [-\mathbf{J}(t') \cdot \frac{\mathbf{E}}{i\omega} e^{-i\omega t'}, \mathbf{J}(t)] \right] | \psi_0 \rangle. \quad (1.14.10)$$

The first term inside the bracket in the RHS is the current due to the absence of an external electric field that can safely be ignored in our case. So only the second term survives. Hence, ignoring the negative sign inside the commutator and writing for components α, β $(\alpha, \beta \in x, y, z)$,

$$\langle J_\alpha(t) \rangle = \frac{1}{\hbar\omega} \int_{-\infty}^{t} dt' \langle \psi_0 | [J_\beta(t'), J_\alpha(t)] | \psi_0 \rangle E_\beta e^{-i\omega t'}. \quad (1.14.11)$$

where $\mathbf{J} \cdot \mathbf{E}$ is written as $J_\beta E_\beta$.

Since the system is invariant under time translation, the above correlation depends on $t - t'$ and not on t and t' individually. Introducing a new variable $t - t' = \tilde{t}$,

$$\langle J_\alpha(t) \rangle = \frac{1}{\hbar\omega} \left(\int_{0}^{\infty} d\tilde{t} \, e^{i\omega\tilde{t}} \langle \psi_0 | [J_\beta(0), J_\alpha(\tilde{t})] | \psi_0 \rangle \right) E_\beta e^{-i\omega t}, \quad (1.14.12)$$

where the term inside the bracket in the RHS can be written as $(\sigma_{\alpha\beta})$, $\sigma_{\alpha\beta}$ being the components of the conductivity tensor. Note that the time t dependence is outside the integral and appears as $e^{-i\omega t}$. Thus, in the linear response regime, if an electric field of frequency ω is applied, the current responds to the external field by oscillating with the same frequency as the external field.

The Hall conductivity is the off-diagonal component and can be computed using

$$\sigma_{xy}(\omega) = \frac{1}{\hbar\omega} \int_{0}^{\infty} dt \, e^{i\omega t} \langle \psi_0 | [J_y(0), J_x(t)] | \psi_0 \rangle. \quad (1.14.13)$$

$J_x(t)$ can be written as

$$J_x(t) = e^{i\mathcal{H}_0 t/\hbar} J_x(0) e^{-i\mathcal{H}_0 t/\hbar} \quad (1.14.14)$$

in the above expression, and using completeness relation of the states, namely $|\psi_n\rangle \langle\psi_n| = 1$, one obtains

$$\sigma_{xy}(\omega) = \frac{1}{\hbar\omega} \int_{0}^{\infty} dt \, e^{i\omega t} \sum_n \langle \psi_0 | J_y | \psi_n \rangle \langle \psi_n | J_x | \psi_0 \rangle e^{i(E_n - E_0)t/\hbar}$$
$$- \langle \psi_0 | J_x | \psi_n \rangle \langle \psi_n | J_y | \psi_0 \rangle e^{i(E_0 - E_n)t/\hbar}. \quad (1.14.15)$$

Now we shall perform the integral over time, t. Let us write $\frac{E_n}{\hbar} = \omega_n$ and introduce $\omega \to \omega + i\epsilon$, where ϵ is the infinitesimal quantity considered for the convergence

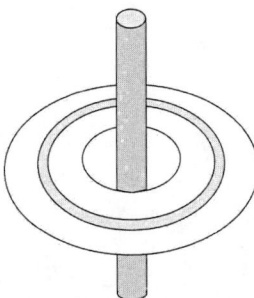

FIGURE 1.12.   The schematic plot shows a disc threaded by a flux perpendicular to the plane and another one in the plane of the disc.

of the integral. Then the Hall conductivity is obtained as

$$\sigma_{xy}(\omega) = -\frac{i}{\omega} \sum_{E_n \neq E_0} \left[\frac{\langle \psi_0| J_y |\psi_n\rangle \langle \psi_n| J_x |\psi_0\rangle}{\hbar\omega + E_n - E_0} - \frac{\langle \psi_0| J_x |\psi_n\rangle \langle \psi_n| J_y |\psi_0\rangle}{\hbar\omega + E_0 - E_n} \right]. \quad (1.14.16)$$

Finally, we shall take the limit $\omega \to 0$ to account for the constant field. For that let us expand the denominator as follows:

$$\frac{1}{\hbar\omega + E_n - E_0} \approx \frac{1}{E_n - E_0} - \frac{\hbar\omega}{(E_n - E_0)^2} + O(\omega^2)$$

$$\frac{1}{\hbar\omega + E_0 - E_n} \approx \frac{1}{E_0 - E_n} - \frac{\hbar\omega}{(E_0 - E_n)^2}. \quad (1.14.17)$$

The first term looks divergent, and such divergence is responsible for the peak in the longitudinal conductivity, which by now we are familiar with. Moreover, $\sigma_{xy}(\omega)$ should not contain a term that is independent of ω. This is the DC conductivity ($\omega = 0$) that is absent in a translationally invariant system. Thus, finally one arrives at

$$\sigma_{xy}(\omega) = i\hbar \sum_{n \neq 0} \frac{\langle \psi_0| J_y |\psi_n\rangle \langle \psi_n| J_x |\psi_0\rangle - \langle \psi_0| J_x |\psi_n\rangle \langle \psi_n| J_y |\psi_0\rangle}{(E_n - E_0)^2}. \quad (1.14.18)$$

This is the Kubo formula for the Hall conductivity.

    To proceed further, let us assume a specific case of perturbing a system below [69]. Consider a quantum Hall sample in the form of a torus (or a donut). Let us thread two fluxes Φ_x and Φ_y (instead of one) as shown in Fig. 1.12. Owing to this, the gauge potentials can be written as

$$A_x = \frac{\Phi_x}{L_x}$$

$$A_y = \frac{\Phi_y}{L_y} + Bx. \tag{1.14.19}$$

It is clear that the states of the quantum system are sensitive to non-integer values of Φ_i/Φ_0 ($i = x, y$), where $\Phi_0 = \frac{h}{e}$. Specifically, if we increase either Φ_x or Φ_y from 0 to Φ_0, then the spectrum of the quantum system must remain invariant. Hence, writing the perturbation Hamiltonian in terms of the fluxes,

$$\mathcal{H}' = -\sum_{i \in 1,2} \frac{J_i \Phi_i}{L_i}. \tag{1.14.20}$$

To the first order corresponding to this perturbation term, the modified ground state becomes

$$|\psi_0'\rangle = |\psi_0\rangle + \sum_{\substack{\psi_n \neq \psi_0 \\ E_n \neq E_0}} \frac{\langle \psi_n| H' |\psi_0\rangle}{E_n - E_0} |\psi_n\rangle. \tag{1.14.21}$$

Thus, writing

$$\Delta |\psi\rangle = |\psi_0'\rangle - |\psi_0\rangle = \sum_{\substack{\psi_n \neq \psi_0 \\ E_n \neq E_0}} \frac{\langle \psi_n| H' |\psi_0\rangle}{E_n - E_0} |\psi_n\rangle. \tag{1.14.22}$$

Considering infinitesimal changes in Φ_i, we can write

$$\left| \frac{\partial |\Delta\psi\rangle}{\partial \Phi_i} \right\rangle = -\frac{1}{L_i} \sum_n \frac{\langle \psi_n| J_i |\psi_0\rangle}{E_n - E_0} |\psi_n\rangle. \tag{1.14.23}$$

Terms like those in the RHS of the above equation appeared in the Hall conductivity. Let us write the total Hall conductivity including the area factor $L_x L_y$ of the sample, which can be written as

$$\sigma_{xy} = i\hbar L_x L_y \sum_{\substack{\psi_n \neq \psi_0 \\ E_n \neq E_0}} \frac{\langle \psi_0| J_y |\psi_n\rangle \langle \psi_n| J_x |\psi_0\rangle - \langle \psi_0| J_x |\psi_n\rangle \langle \psi_n| J_y |\psi_0\rangle}{(E_n - E_0)^2}$$

$$= i\hbar \left[\left\langle \frac{\partial\psi_0}{\partial\Phi_y} \middle| \frac{\partial\psi_0}{\partial\Phi_x} \right\rangle - \left\langle \frac{\partial\psi_0}{\partial\Phi_x} \middle| \frac{\partial\psi_0}{\partial\Phi_y} \right\rangle \right] \tag{1.14.24}$$

$$= i\hbar \left[\frac{\partial}{\partial\Phi_y} \left\langle \psi_0 \middle| \frac{\partial\psi_0}{\partial\Phi_x} \right\rangle - \frac{\partial}{\partial\Phi_x} \left\langle \psi_0 \middle| \frac{\partial\psi_0}{\partial\Phi_y} \right\rangle \right]. \tag{1.14.25}$$

$\left\langle \psi_0 \middle| \frac{\partial\psi_0}{\partial\Phi_x \partial\Phi_y} \right\rangle$ will cancel from both the terms. This is the Kubo formula for Hall conductivity.

2

Symmetry and Topology

In this chapter, we shall discuss the interplay of symmetry and topology that are essential in understanding the topological protection rendered by the inherent symmetries and how the topological invariants are related to physical quantities.

2.1   Introduction

Point set topology is a disease from which the human race will soon recover.

— H. Poincaré (1908)

Poincaré conjecture was the first conjecture made on topology which asserts that a three-dimensional (3D) manifold is equivalent to a sphere in 3D subject to the fulfilment of a certain algebraic condition of the form $f(x, y, z) = 0$, where x, y and z are complex numbers. G. Perelman has (arguably) solved the conjecture in 2006 [4]. However, on practical aspects, just the reverse of what Poincaré had predicted happened. Topology and its relevance to condensed matter physics have emerged in a big way in recent times. The 2016 *Nobel Prize* awarded to D. J. Thouless, J. M. Kosterlitz, F. D. M. Haldane and C. L. Kane and E. Mele getting the *Breakthrough Prize* for contribution to fundamental physics in 2019 bear testimony to that.

Topology and geometry are related, but they have a profound difference. Geometry can differentiate between a square from a circle, or between a triangle and a rhombus; however, topology cannot distinguish between them. All it can say is that individually all these shapes are connected by continuous lines and hence are identical. However, topology indeed refers to the study of geometric shapes where the focus is on how properties of objects change under continuous deformation, such as stretching and bending; however, tearing or puncturing is not allowed. The objective is to determine whether such a continuous deformation can lead to a change from one geometric shape to another. The connection to a problem of deformation of geometrical shapes in condensed matter physics may be established if the Hamiltonian for a particular system can be continuously transformed via tuning of one (or more) of the parameter(s) that the Hamiltonian depends on. Should there be no change in the number of energy modes below the Fermi energy during the process of transformation, then the two systems (that is, before and after the transformation) belong to the same topology class. In the process, *something* remains invariant. If that *something* does not remain invariant, then there occurs a topological phase transition. This phase transition can occur from one topological phase to another or from a topological phase to a trivial phase.

In the following discussion, we present the geometric aspects of topology and relate the integral of the geometric properties over closed surfaces to the topological invariants. It turns out that the *geometric property* and the *closed surface* have smooth connection to physical observables. As we shall see soon, in 1982, Thouless, Kohmoto, Nightingale and den Nijs [70] linked the topological invariant to the quantized Hall conductivity.

To test many of the concepts that we are going to discuss in this chapter, we choose two prototype systems, one each in one (1D) and two dimensions (2D). In 1D, we consider a tight-binding model, with dimerized hopping, and in 2D, we consider graphene, which has been a hobby horse even several years before its experimental discovery. The theme is to discuss the interplay of symmetry and the topological properties. Particularly, in 2D an important highlight in this direction is put forward by Haldane [6], who had proposed a non-trivial topological phase by breaking one of the fundamental symmetries, namely the time reversal symmetry. Finally, after the experimental discovery of graphene, yet another distinct topological state of matter was discovered by Kane and Mele [7], which has culminated into an emerging field of spintronics.

We have seen earlier that the Hall conductivity (or the resistivity) is quantized in the unit of e^2/h (or h/e^2) within a splendid precision, so much so that the quantity h/e^2 can define the standard of resistance (= 25.5 kΩ). Clearly, the quantization is independent of the details of the Hamiltonian, for example, the nature of the sample, the strength of the magnetic field and disorder present in the system. It is realized later that the universality of the phenomenon arises due to 'topological' protection of the energy modes that exist at the edges of a quantum Hall sample, which possess completely different character as compared to the ones that exist in the bulk of the sample. Thus, an understanding emerges that says that a physical observable (which is either the resistivity or the conductivity) can be represented mathematically by a topological invariant. This invariant does not change even when the Hamiltonian changes (for example, when the strength of the magnetic field is varied), until and unless a phase transition occurs that will show up via an abrupt change in the value of the topological invariant. There is an elegant explanation of the physics involved with such a universal phenomenon, which brings us to the subject of topology.

Topology in its usual sense deals with the geometry of the objects; in the same spirit, here we shall study the geometrical properties of the Hilbert space for the system under consideration. The ideas are best demonstrated for a quantum Hall system that undergoes a series of transitions from a conducting to an insulating state as a function of the external magnetic field. In the process, the topological invariant, for example, the Chern number in this case (we shall discuss this later), jumps from one integral value to another. Thus, the system repeatedly undergoes through a series of topological phase transitions. In the following discussion, we describe this topological phase transition in more general terms.

Consider two Hamiltonians, $\mathcal{H}_1$ and $\mathcal{H}_2$, both of which are functions of a tunable parameter, say, β. If the corresponding energy spectra $\epsilon_1(\beta)$ and $\epsilon_2(\beta)$ are such that the number of energy levels below the zero energy (zero energy is usually the Fermi energy) always remains same for all values of β, then the Hamiltonians can be continuously transformed (or deformed as we see the analogy of a cup and donut later), and there is no phase transition. Now, consider either $\mathcal{H}_1$ or $\mathcal{H}_2$. If, for either of them, the spectrum is such that the number of energy levels vary as a function of β, that is, if any (or more) level crosses the zero energy, then the 'invariant' changes (from one integer value to another), and one encounters the system is undergoing from one topological phase to another. A quantum Hall system shows a similar transition, where the Hall conductivity changes from ne^2/h to $(n+1)e^2/h$, where n is strictly an integer.

Thus, the study of topology deals with objects (or Hamiltonians) that can be continuously transformed (or deformed) from one to another without puncturing or tearing the object (or without even closing the energy gap for the quantum system). For geometrical objects, being able to continuously transform depends on the number of 'holes' or 'genus' that are preserved during the course of the transformation. For example, a soccer ball can be deformed smoothly into a wine glass since both of them have no hole (zero genus), while a mug (as shown in Fig. 2.1) can be transformed smoothly into a donut with one hole (genus equal to 1). The first case with zero hole is called 'topologically trivial', and the second with finite number of holes (one in this case) is termed as 'topologically non-trivial'.

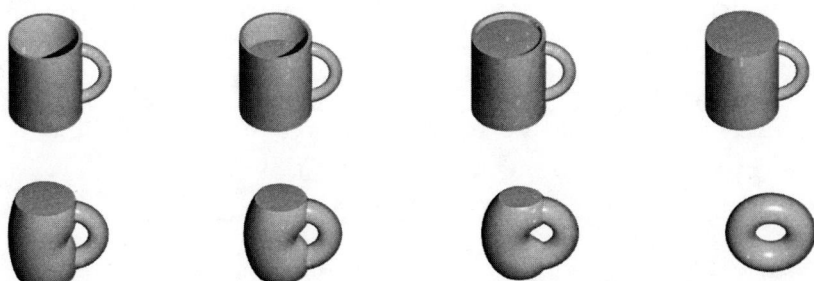

FIGURE 2.1.   A mug can be transformed smoothly into a donut. The handle of the mug remains invariant and emerges as 'hole' of the donut. Thus, the mug and the donut belong to the same universality class.

2.2   Gauss–Bonnet theorem

The Gauss–Bonnet theorem in differential geometry is about evaluation of the surface integral of a Gaussian surface. Here we state the theorem without proof. In the most general form, for a closed polyhedral surface, the theorem can be stated as

$$\int_{\partial R} k_g(s)ds + \iint_R KdA = 2\pi\chi(R), \qquad (2.2.1)$$

where R denotes a regular region with the boundary ∂R of R, K is the Gaussian curvature, s denotes the arc length of the curves. Further, $\chi(R)$ is called the Euler–Poincare characteristic. The first term on the left is the integral of the Gaussian curvature over the surface; the second one is the integral of the geodesic curvature of the boundary of the surface. Thus, the Gauss–Bonnet theorem simply states that the total curvature of R plus the total geodesic curvature of ∂R is a constant.

As an example, we consider the simplest case, that is a sphere of radius R. The Gaussian curvature[1] is $1/r^2$ and the corresponding area is

$$\iint_R KdA = K \times \text{Area} = \frac{1}{r^2} \times 4\pi r^2 = 4\pi. \qquad (2.2.2)$$

Again,

$$\iint_R KdA = 2\pi\chi(R). \qquad (2.2.3)$$

---

[1] For a geometry with two different radii of curvature, such as a convex lens, the Gaussian curvature is $\frac{1}{r_1 r_2}$.

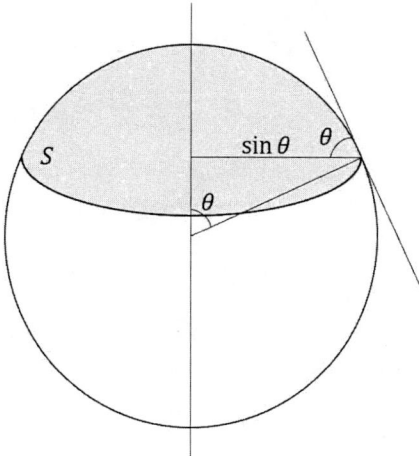

FIGURE 2.2. A sphere with a polar cap. The figure will aid in calculating the area S of the polar cap.

Thus,

$$\chi(R) = 2.$$

Thus, the Euler–Poincare characteristic of a sphere is 2. Suppose we wish to extend this argument to other closed, however, not necessarily convex surfaces in a 3D space. For that, consider the polar cap of unit radius (see Fig. 2.2). The area is given by

$$S = \int_0^{\theta} 2\pi \sin\theta' d\theta' = 2\pi(1 - \cos\theta). \qquad (2.2.4)$$

Thus,

$$\int_R KdA = 1 \times (\text{Area of } S) = 2\pi(1 - \cos\theta).$$

The geodesic curvature K is $1/\tan\theta$. Thus,

$$\int_S K_g ds = K_g \times \text{length}(S) = \frac{1}{\tan\theta}. \qquad (2.2.5)$$

Hence,

$$\int_S K_g ds + \int_R KdA = 2\pi(1 - \cos\theta) + 2\pi\cos\theta = 2\pi = 2\pi\chi(R),$$

thereby yielding

$$\chi(R) = 1.$$

In fact, an alternate form for the Gauss–Bonnet theorem is more useful for our purpose, which states that, for a closed convex surface, the integral over the

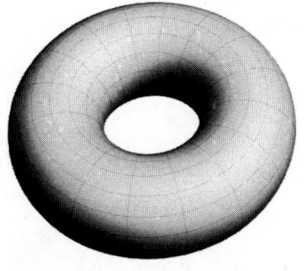 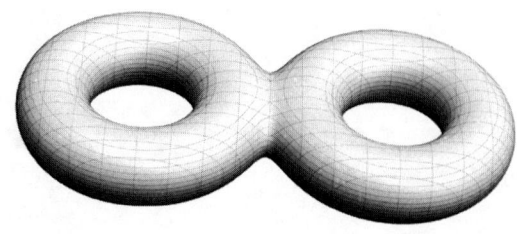

FIGURE 2.3.   (Left) A donut with genus (or a hole) equal to 1. (Right) A two-hole object has genus equal to 2.

Gaussian curvature can be expressed in terms of the number of holes or the genus of the surface. Thus, a simplified (and more relatable for us) version reads

$$\iint K dA = 2\pi(2 - 2g). \tag{2.2.6}$$

Since a sphere has no holes ($g = 0$), the integral of the curvature yields

$$\iint K dA = 4\pi, \tag{2.2.7}$$

a result that we have seen earlier. Let us look at a case where the genus is non-zero ($g \neq 0$), such as a torus that is topologically equivalent to a mug as we have seen earlier.

For a torus, the Euler–Poincare characteristic has a value zero. This implies that irrespective of how we bend or deform it, the integrated curvature vanishes. Refer to Fig. 2.3 (left-hand panel), where there is a positive curvature on the outer surface and a negative curvature on the inner surface, thereby resulting in a zero total curvature. This is consistent with the Gauss–Bonnet theorem, which states that the integral of the Gaussian curvature is $2\pi(2 - 2g)$. Since $g = 1$ here, the integral is zero and so the Euler–Poincare characteristic $\chi(R)(= 2 - 2g)$ is zero as well. Similarly, a two-holed donut (see right-hand panel of Fig. 2.3) will have $\chi(R) = -2$, and hence a negative integrated Gaussian curvature.

Based on the preceding discussion, a sketchy idea emerges on the relationship between topology and the properties of quantum systems. However, it still remains unclear how the ideas can relate to the properties of materials. At the moment let us talk about crystalline solids for which electron wavefunction is given by Bloch's theorem, namely

$$\psi(\mathbf{r}) = e^{i\mathbf{k}\cdot\mathbf{r}} u_{\mathbf{k}}(\mathbf{r}), \tag{2.2.8}$$

where the periodicity of the crystal potential, that is, $V(\mathbf{r}) = V(\mathbf{r} + \mathbf{R})$ is captured by the amplitude function $u_\mathbf{k}(\mathbf{r})$, such that

$$u_\mathbf{k}(\mathbf{r} + \mathbf{R}) = u_\mathbf{k}(\mathbf{r}), \tag{2.2.9}$$

where $\mathbf{k}$ denotes the crystal momentum and is distinct than the usual momentum $(= -i\hbar\nabla)$ [63]. The crystal momentum is restricted within the first Brillouin zone (BZ), where the latter is a region in the k-space with periodic boundaries. As the crystal momentum is varied, we map out the energy bands, and one obtains the band structure. The BZ plays the role of the surface over which the integral of the Gaussian curvature is taken, which we have discussed earlier.

Now that brings us to the question: *what is the analogue of the Gaussian curvature for a crystalline solid?* To understand this, consider the (non-degenerate) ground state of a Hamiltonian that depends upon a number of parameters that are time dependent. The adiabatic theorem states that if the Hamiltonian is now changed slowly[2] with respect to the parameters, the system remains in its time-dependent ground state. However, there is something more to it. As the ground state is evolved in time, in addition to the trivial dynamical phase, there may emerge an irreducible geometric phase that comes into play, namely the Berry phase put forward by M. V. Berry in 1984 [9].

In the following, we discuss the origin of the Berry phase and the Berry curvature, which is analogous to the Gaussian curvature. The integral of the Berry curvature over the BZ will be shown to yield a constant (or more appropriately an invariant) known as the Chern number, which is analogous to the RHS of Eq. (2.2.6) or the Euler–Poincare characteristic.

2.3 Berry phase

Consider a particle is in the ground state of a box of length L. Suppose the box slowly expands such that $L(t)$ is a slow function of time. The adiabatic principle says that if the expansion is slow, then the particle always remains in the ground state at any time t. It is true for any state of the system. More generally, consider a Hamiltonian $\mathcal{H}(\lambda(t))$, where λ is a parameter that changes slowly. Now the adiabatic principle says that if the particle starts out in the nth eigenstate of $\mathcal{H}(\lambda(0))$, it will land nth instantaneous eigenstate of $\mathcal{H}(\lambda(t))$ at a time t.

---

[2]The time scale for change is larger than the inverse level spacing (level spacing implies difference between subsequent energy levels) of the system.

The question is what is the solution of the Schrödinger equation in this approximation? A reasonable guess is

$$|\psi(t)\rangle = \exp\left(-\frac{i}{\hbar}\int \varepsilon_n(t')dt'\right)|\phi_n(t)\rangle,$$ (2.3.1)

where

$$\mathcal{H}(t)|\phi_n(t)\rangle = \varepsilon_n(t)|\phi_n(t)\rangle.$$ (2.3.2)

If $\mathcal{H}$ does not vary with time, then clearly the phase is correct. However, it is not so in the case $\mathcal{H}$ depends on time.

To see what is missing in the above ansatz, let us modify it slightly.

$$|\psi(t)\rangle = C(t)\exp\left[-\frac{i}{\hbar}\int_0^t \varepsilon_n(t')dt'\right]|\phi_n(t)\rangle.$$ (2.3.3)

In Eq. (2.3.1), $C(t) = 1$. Applying the Schrödinger equation $(i\hbar\frac{\partial}{\partial t} - \mathcal{H})$ to Eq. (2.3.3) and simplifying, for the time dependence of C, one gets

$$\dot{C}(t) = -C(t)\langle\phi_n(t)|\frac{d}{dt}|\phi_n(t)\rangle.$$ (2.3.4)

This yields a solution of the form

$$C(t) = C(0)\exp\underbrace{\left[-\int_0^t \langle\phi_n(t')|\frac{d}{dt'}|\phi_n(t')\rangle\,dt'\right]}_{e^{i\gamma}}$$

$$= C(0)e^{i\gamma},$$ (2.3.5)

where

$$\gamma = i\int_0^t \langle\phi_n(t')|\frac{d}{dt'}|\phi_n(t')\rangle\,dt'.$$ (2.3.6)

This extra phase is called the Berry phase or the geometric phase. It is also called the Berry–Pancharatnam phase, and is quite a familiar quantity in the field of optics.

In general, phases do not give rise to measurable consequences, since the eigenstates are defined only up to a phase factor. Even here, it may be thought that we can define new Berry states to absorb the phase, namely

$$\left|\phi_n'(t)\right\rangle = e^{i\chi(t)}\left|\phi_n(t)\right\rangle, \tag{2.3.7}$$

then

$$i\left\langle\phi_n'(t)\right|\frac{d}{dt}\left|\phi_n'(t)\right\rangle = i\left\langle\phi_n(t)\right|\frac{d}{dt}\left|\phi_n(t)\right\rangle - \frac{d\chi}{dt}. \tag{2.3.8}$$

Now suppose the parameter $\chi(t)$ changes the Hamiltonian in such a manner that after a complex cycle

$$\mathcal{H}(0) = \mathcal{H}(t = T).$$

The end result is

$$i\oint\left\langle\phi_n'(t)\right|\frac{d}{dt}\left|\phi_n'(t)\right\rangle = i\oint\left\langle\phi_n(t)\right|\frac{d}{dt}\left|\phi_n(t)\right\rangle - (\chi(T) - \chi(0)). \tag{2.3.9}$$

The last term on the RHS is an irreducible phase that does not cancel under redefinition of the Berry states. Remember, χ arises from γ, which we denote as the Berry phase. Single valuedness of χ demands

$$\chi(T) - \chi(0) = 2\pi n, \quad \text{where } n \text{ is an integer.}$$

In the Gauss–Bonnet theorem, just like an object with a genus '1' cannot be smoothly transformed into another with genus 'zero' or '2' (unless, of course, something drastic, that is, tearing or puncturing is done to the object), a system with a non-zero Chern number cannot be transformed into that with zero Chern number.

2.4 Hall conductivity and the Chern number

Remember, the spectrum of the Hamiltonian depends upon Φ_i mod Φ_0.[3] If there is a remainder, then the division does not yield an integer, and if there is none, then $\Phi_i/\Phi_0 = $ integer. Φ_i being parameters of $\mathcal{H}'$, Φ_i's are periodic functions. To emphasize the periodicity, we shall introduce angular variables θ_i such that

$$\theta_i = \frac{2\pi\Phi_i}{\Phi_0}, \quad \text{where } \theta_i \in [0, 2\pi]. \tag{2.4.1}$$

As θ_i increases from 0 to 2π, Φ_i increases from $0 \to \Phi_0$. Now rewrite $\frac{\partial}{\partial\Phi_i}$ as $\frac{\partial}{\partial\theta_i}$ and introduce a quantity called the Berry connection, which is defined on the surface

---

[3] Φ_i *mod* Φ_0 refers to the remainder when Φ_i is divided by Φ_0.

of the torus as

$$\mathcal{A}_i(\Phi) = -i \langle \psi_0 | \frac{\partial}{\partial \theta_i} | \psi_0 \rangle. \tag{2.4.2}$$

Further, we define a quantity called the Berry curvature (analogous to the magnetic field),

$$\mathcal{F}_{xy} = \frac{\partial \mathcal{A}_x}{\partial \theta_y} - \frac{\partial \mathcal{A}_y}{\partial \theta_x} = (\nabla_\theta \times \mathcal{A}) = -i \left[\frac{\partial}{\partial \theta_y} \left\langle \psi_0 \left| \frac{\partial \psi_0}{\partial \theta_x} \right\rangle - \frac{\partial}{\partial \theta_x} \left\langle \psi_0 \left| \frac{\partial \psi_0}{\partial \theta_y} \right\rangle \right]. \right.$$
$$\tag{2.4.3}$$

Note that the last term in Eq. (2.4.3) is the Hall conductivity, which is formally written in terms of the Berry curvature as

$$\sigma_{xy} = -\frac{e^2}{h} \mathcal{F}_{xy}. \tag{2.4.4}$$

We are still left with the task of understanding the quantization of σ_{xy}, which is central to the discussion on quantum Hall effect. We can now integrate over the surface of the torus to get the total conductivity,

$$\sigma_{xy} = -\frac{e^2}{h} \int_{\text{torus}} \frac{d^2\theta}{2\pi} \mathcal{F}_{xy}. \tag{2.4.5}$$

The quantity $C = \frac{1}{2\pi} \int d^2\theta \, \mathcal{F}_{xy}$ is called the first Chern number. Thus, if we average over the fluxes, the conductivity assumes a form

$$\sigma_{xy} = -\frac{e^2}{h} C, \tag{2.4.6}$$

C is necessarily an integer. It is also referred to as the TKNN invariant after Thouless, Kohmoto, Nightingale and den Nijs [70].

Here we provide an argument that the Chern number, which is the integral over the Berry curvature, is indeed an integer. For simplicity, let us assume a translationally invariant system in which the eigenstates can be represented by the Bloch functions. That is, $|\psi_0\rangle$ appearing above can be written as $u_k e^{i\phi_k}$, where u_k captures the periodicity of the lattice. The Berry connection requires a derivative to be taken, namely $\langle \psi_0 | \frac{\partial}{\partial \theta} | \psi_0 \rangle$, which is equal to

$$\frac{\partial}{\partial \phi_k}(u_k e^{i\phi_k}) \approx \nabla_k \phi_k. \tag{2.4.7}$$

Now when one takes an integral over the BZ (which is equivalent to the surface of the torus in real space), then $(\nabla_k \phi_k) \cdot dk$ over a closed surface is zero.

$$\oint \nabla_k \cdot \phi_k dk = 0. \tag{2.4.8}$$

This says that the measurable quantity $e^{i\phi}$ obeys[4]

$$e^{i\phi(0)} = e^{i\phi(2\pi)}. \tag{2.4.9}$$

Thus,

$$|\phi(0) - \phi(2\pi)| = 2\pi \times \text{(some integer)}$$
$$= 2\pi C. \tag{2.4.10}$$

Hence, the integral over the curvature

$$\oint \mathcal{F} \frac{d^2 k}{2\pi} = \oint \nabla_k \phi_k dk = \frac{2\pi C}{2\pi} = C. \tag{2.4.11}$$

Thus, Chern number is an integer, and hence we get the Hall conductivity to be quantized in unit of e^2/h. The above calculations are of course applicable to IQHE.

In a quantum Hall system, the Hall conductivity is given by

$$\sigma_{xy} = Ce^2/h = ne^2/h, \quad \text{where } C = \text{ Chern number.}$$

It can be argued that the Chern number is always an integer. Further, the Berry curvature $\mathcal{F}$ is defined as curl of the Berry connection, namely

$$\mathcal{F} = \nabla \times \mathcal{A}. \tag{2.4.12}$$

$\mathcal{F}$ is analogous to the magnetic field. The Chern number is defined as the surface integral of the Berry curvature over a surface enclosed.

---

[4]It should be remembered that ϕ is not a measurable quantity, while $e^{i\phi}$ is a measurable quantity.

2.5   Discrete symmetries

To elucidate more on the topological invariance in materials, we discuss a few discrete symmetries of the Hamiltonian and how they interplay with the topological properties. In this context, we wish to discuss three symmetries, namely the inversion symmetry (also dubbed as sublattice symmetry in certain cases) and the time reversal symmetry. A third symmetry that we shall talk about is the contribution of the above two and is known as the particle–hole symmetry. We shall discuss these symmetries one by one.

2.5.1   Inversion symmetry

Let us consider an eigenstate in the position basis $|\psi(\mathbf{r})\rangle$, so that

$$\mathbf{r}\,|\psi(\mathbf{r})\rangle = |\mathbf{r}|\,|\psi(\mathbf{r})\rangle. \qquad (2.5.1.1)$$

Now we define the inversion symmetry or the parity operator $\mathcal{P}$ such that

$$\mathcal{P}\,|\psi(\mathbf{r})\rangle = |\psi(-\mathbf{r})\rangle. \qquad (2.5.1.2)$$

Now,

$$\mathbf{r}\mathcal{P}\,|\psi(\mathbf{r})\rangle = \mathbf{r}\,|\psi(-\mathbf{r})\rangle = -|\mathbf{r}|\,|\psi(-\mathbf{r})\rangle. \qquad (2.5.1.3)$$

If we act $\mathcal{P}^\dagger$ on both sides (remembering $\mathcal{P}^\dagger = \mathcal{P}^{-1} = \mathcal{P}$),

$$\begin{aligned}
\mathcal{P}^\dagger\,\mathbf{r}\,\mathcal{P}\,|\psi(\mathbf{r})\rangle &= -|\mathbf{r}|\mathcal{P}^\dagger\,|\psi(-\mathbf{r})\rangle \\
&= -|\mathbf{r}|\mathcal{P}\,|\psi(-\mathbf{r})\rangle \\
&= -|\mathbf{r}|\,|\psi(\mathbf{r})\rangle = -\mathbf{r}\,|\psi(\mathbf{r})\rangle.
\end{aligned} \qquad (2.5.1.4)$$

Thus,

$$\mathcal{P}^\dagger\,\mathbf{r}\,\mathcal{P} = -\mathbf{r}. \qquad (2.5.1.5)$$

This yields

$$\mathcal{P}^\dagger\,\mathbf{r} = -\mathbf{r}\mathcal{P} \qquad (2.5.1.6)$$

$$\text{or, } \{\mathcal{P},\mathbf{r}\} = 0. \qquad (2.5.1.7)$$

Hence, the parity operator anticommutes with the position operator.

Let us now explore the analogous scenario for the momentum operator. For this purpose, it is convenient to introduce the transformation operator $T(\mathbf{a})$[5] that translates a state $|\psi(\mathbf{r})\rangle$ to $|\psi(\mathbf{r}+\mathbf{a})\rangle$, where $\mathbf{a}$ denotes a fixed length; for example, $\mathbf{a}$ can be the lattice constant. That is,

$$T(\mathbf{a})|\psi(\mathbf{r})\rangle = |\psi(\mathbf{r}+\mathbf{a})\rangle$$

$$\mathcal{P}^\dagger T(\mathbf{a})\mathcal{P}|\psi(\mathbf{r})\rangle = T(-\mathbf{a})|\psi(\mathbf{r})\rangle, \qquad (2.5.1.8)$$

which yields

$$\mathcal{P}^\dagger\, T(\mathbf{a})\, \mathcal{P} = T(-\mathbf{a}). \qquad (2.5.1.9)$$

This demands that the translation operator is of the form

$$T(\mathbf{a}) = e^{i\,\mathbf{k}\cdot\mathbf{a}}. \qquad (2.5.1.10)$$

Expanding for infinitesimal translations,

$$\mathcal{P}^\dagger\, \mathbf{p}\, \mathcal{P} = -\mathbf{p}. \qquad (2.5.1.11)$$

Thus, similar to the position operator, the momentum operator too anticommutes with the parity operator.

Since both $\mathbf{r}$ and $\mathbf{p}$ anticommute, the angular momentum $\mathbf{L}\,(=\mathbf{r}\times\mathbf{p})$ commutes with $\mathcal{P}$. In a 3D orthogonal coordinate system, one can invert it about any of the axes. For example, in a Cartesian coordinate system,

(i) an inversion about the z-axis is denoted as $\sigma_h(xy)$,
(ii) about the y-axis it is $\sigma_v(xz)$ and
(iii) $\sigma_v(yz)$ denotes the inversion about the x-axis.

Here σ denotes an inversion operation and has got nothing to do with the Pauli matrices. Under these operations, the position and the angular momentum variable transforms as

(i) $\sigma_h(xy): x \to x, y \to y, z \to -z$
   $L_x \to -L_x, L_y \to -L_y, L_z \to L_z$
(ii) $\sigma_v(xz): x \to x, y \to -y, z \to z$
   $L_x \to -L_x, L_y \to L_y, L_z \to -L_z$
(iii) $\sigma_v(yz): x \to -x, y \to y, z \to z$
   $L_x \to L_x, L_y \to -L_y, L_z \to -L_z$

---

[5]Distinguish between $T(a)$ for the translation operator and $\mathcal{T}$ for the time reversal operator.

2.5.2 Time reversal symmetry

Now we shall discuss time reversal symmetry. It is obvious that under time reversal, the time variable t changes to $-t$. This makes the position ($\mathbf{r}(t)$) and the momentum variable ($\mathbf{p}(t)$) transform under time reversal as $\mathbf{r}(-t)$ and $-\mathbf{p}(t)$, respectively. The angular momentum $\mathbf{L}(t)(= \mathbf{r} \times \mathbf{p})$ thus also becomes $-\mathbf{L}(-t)$ under time reversal. Similar outcomes are expected when $\mathbf{r}(t)$, $\mathbf{p}(t)$ and $\mathbf{L}(t)$ are quantum mechanical operators. Additional inputs to the ongoing discussion can be received from the behaviour of the electric field $\mathbf{E}(\mathbf{r}, t)$ and the magnetic field $\mathbf{B}(\mathbf{r}, t)$ vectors under time reversal. $\mathbf{E}(\mathbf{r}, t)$ does not change sign under time reversal (refer to the Maxwell's equations, $\nabla \cdot \mathbf{E} = \frac{\rho}{\epsilon_0}$, where charge density $\rho(\mathbf{r})$ does not change sign); however, $\mathbf{B}(\mathbf{r}, t)$ changes sign (owing to $\nabla \times \mathbf{B} = \mu_0 \mathbf{J}$, where $\mathbf{J}(\mathbf{r}, t)$ is the current density, and it changes sign under time reversal).

Now, consider a quantum state $\psi(t)$ that obeys the Schrödinger equation

$$i\hbar \frac{\partial \psi(\mathbf{r}, t)}{\partial t} = \mathcal{H}\psi(\mathbf{r}, t). \tag{2.5.2.1}$$

In the following discussion, we suppress the $\mathbf{r}$ dependence of ψ, and simply write $\psi(t)$ which upon the application of the time reversal operator yields $\psi'(-t)$. Mathematically,

$$\mathcal{T}\left|\psi(t)\right\rangle = \left|\psi'(-t)\right\rangle. \tag{2.5.2.2}$$

In order to find $\psi'(-t)$, let us look at the solution of Eq.(2.5.2.1),

$$\left|\psi(t)\right\rangle = e^{-i\mathcal{H}t/\hbar}\left|\psi(0)\right\rangle. \tag{2.5.2.3}$$

For $t = 0$, apply the time reversal operator, that is, $\mathcal{T}\left|\psi(0)\right\rangle$. Now, let it evolve forward in time, which means we get a state

$$e^{-i\mathcal{H}t/\hbar}\mathcal{T}\left|\psi(0)\right\rangle.$$

For the Hamiltonian to be invariant under time reversal, this state should be the same as $\mathcal{T}\psi(-t)$, which is equivalent to

$$\mathcal{T}e^{i\mathcal{H}t/\hbar}\left|\psi(0)\right\rangle.$$

Thus,

$$\mathcal{T}e^{i\mathcal{H}t/\hbar}\left|\psi(0)\right\rangle = e^{-i\mathcal{H}t/\hbar}\mathcal{T}\left|\psi(0)\right\rangle.$$

For small time δt, we can expand the exponential and write

$$\mathcal{T}i\mathcal{H} = -i\mathcal{H}\mathcal{T}. \tag{2.5.2.4}$$

A natural intuition (albeit wrong as shown later) is to cancel the 'i' from both sides of Eq. (2.5.2.4). This yields

$$\mathcal{T}\mathcal{H} = -\mathcal{H}\mathcal{T}, \tag{2.5.2.5}$$

which implies

$$\mathcal{T}\mathcal{H} + \mathcal{H}\mathcal{T} = 0 \quad \text{or} \quad \{\mathcal{T}, \mathcal{H}\} = 0. \tag{2.5.2.6}$$

But that cannot be correct, since we have assumed that the time reversal operation to be a valid symmetry operation. This means that cancelling the 'i' from both sides in Eq. (2.5.2.4) was not a legitimate step.

Reconciliation is possible if we understand that time reversal, unlike most other operations in quantum mechanics, is an anti-unitary operation. To remind ourselves a unitary operation U satisfies $UU^\dagger = \mathbb{1}$ or a unitary operator acting on a state $\alpha \,|\psi\rangle$ yields

$$U(\alpha \,|\psi\rangle) = \alpha U \,|\psi\rangle. \tag{2.5.2.7}$$

This is also the property of a linear operator. However, for an antilinear operator A, one gets

$$A(\alpha \,|\psi\rangle) = \alpha^* A \,|\psi\rangle, \tag{2.5.2.8}$$

which means that the antilinear operator does a complex conjugation. This resolves the dilemma caused by the naive cancellation of 'i' in Eq. (2.5.2.4). Thus, the factor 'i' on the LHS of Eq. (2.5.2.4) is complex conjugated when it encounters $\mathcal{T}$ on the way pulling it. This yields an extra minus sign, which cancels with the one on the RHS yielding

$$[\mathcal{T}, \mathcal{H}] = 0. \tag{2.5.2.9}$$

This is familiar with the notion that $\mathcal{H}$ is the time reversal invariant (we have to deliberately break time reversal invariance of the Hamiltonian, either via an external magnetic field or some other means), and hence the Hamiltonian should commute with the time reversal operator.

Any anti-unitary operator can be written as a product of a unitary operator U multiplied by a complex conjugation operator $\mathcal{K}$ such that

$$\mathcal{T} = U\mathcal{K}.$$

A special case in this regard deserves a mention, that is, the case for an $s = \frac{1}{2}$ particle. Spin being an angular momentum, it is odd under time reversal, that is,

$$\mathcal{T}\sigma\mathcal{T}^{-1} = -\sigma. \tag{2.5.2.10}$$

This implies $\mathcal{K}\sigma_x\mathcal{K}^{-1} = \sigma_x$, $\mathcal{K}\sigma_y\mathcal{K}^{-1} = -\sigma_y$ and $\mathcal{K}\sigma_z\mathcal{K}^{-1} = \sigma_z$, which is reasonable as σ_y contains imaginary entries. Thus, for the unitary operator, $U\sigma_x U^{-1} = -\sigma_x$, $U\sigma_y U^{-1} = \sigma_y$ and $U\sigma_z U^{-1} = -\sigma_z$. Thus, the unitary operator U commutes with σ_y, but anticommutes with σ_x and σ_z.

Finally, the form of the time reversal operator for a Hamiltonian corresponding to an $S = \frac{1}{2}$ system is given by (without proof, readers are encouraged to try using $\mathcal{T} = \mathcal{K}e^{-\frac{\pi}{2}\sigma_y/\hbar}$)

$$\mathcal{T} = -i\sigma_y\mathcal{K}. \tag{2.5.2.11}$$

Further, for spinless particles (or integer spin), $\mathcal{T}^2 = 1$, while for $S = \frac{1}{2}$ particles, $\mathcal{T}^2 = -1$.

Let us consider time reversal of the type $\mathcal{T}^2 = -1$. Take a pair of Kramer's degenerate time-reversed states, namely $|\phi_1\rangle$ and $|\phi_2\rangle$, such that $\phi_1 = \mathcal{T}|\phi_2\rangle$. Then,

$$\langle\mathcal{T}\phi_1|\mathcal{T}|\phi_2\rangle = (U|\phi_1^*\rangle)^\dagger(U|\phi_2^*\rangle) \tag{2.5.2.12}$$
$$= |\phi_1^*\rangle^\dagger U^\dagger U|\phi_2^*\rangle = \langle\phi_1|\phi_2\rangle^*.$$

Since they are time-reversed pairs,

$$\langle\mathcal{T}\phi_2|\phi_2\rangle^* = \langle\mathcal{T}^2|\phi_2|\phi_2\rangle \tag{2.5.2.13}$$
$$= \langle\pm\phi_2|\mathcal{T}|\phi_2\rangle = \pm\langle\mathcal{T}\phi_2|\phi_2\rangle^*$$

The $\pm$ sign denotes the eigenvalue of $\mathcal{T}^2$. If $\mathcal{T}^2 = +1$, there is no special information; however, corresponding to $\mathcal{T}^2 = -1$, one obtains

$$\langle\mathcal{T}\phi_2|\phi_2\rangle = 0. \tag{2.5.2.14}$$

This implies that for every eigenstate, there exists a time-reversed partner with the same energy and that is orthogonal to it. This is known as Kramer's degeneracy.

For a bulk Hamiltonian written in momentum space $\mathcal{H}(\mathbf{k})$, the time reversal symmetry reads as

$$\mathcal{H}(\mathbf{k}) = U\mathcal{H}(-\mathbf{k})^* U^\dagger. \tag{2.5.2.15}$$

A consequence of this is the invariance of the Hamiltonian under $\mathbf{k} \to -\mathbf{k}$ in the first BZ. For a Bloch state, $|\psi(\mathbf{k})\rangle$,

$$\mathcal{H}(\mathbf{k})|\psi(\mathbf{k})\rangle = E(\mathbf{k})|\psi(\mathbf{k})\rangle. \tag{2.5.2.16}$$

For the time reversal symmetric Hamiltonian,

$$U\mathcal{H}(-\mathbf{k})^* U^\dagger |\psi(\mathbf{k})\rangle = E(\mathbf{k})|\psi(\mathbf{k})\rangle. \tag{2.5.2.17}$$

If we left multiply by $U^\dagger$ and take the complex conjugate,

$$\mathcal{H}(-\mathbf{k})U^T|\psi(\mathbf{k})\rangle^* = E(\mathbf{k})U^T|\psi(\mathbf{k})\rangle^*, \tag{2.5.2.18}$$

which says that for each $\psi(\mathbf{k})$ there is a time-reversed partner $U^T|\psi(\mathbf{k})\rangle^*$ of $\mathcal{H}(-\mathbf{k})$ that possesses the same energy. Thus, the energy eigenvalues are symmetric with $\mathbf{k} \to -\mathbf{k}$ in the BZ, that is,

$$E(\mathbf{k}) = E(-\mathbf{k}). \tag{2.5.2.19}$$

In Chapter 5, we shall encounter situations where calculation of the topological invariant is simplified with special points being present in the BZ that are connected by time reversal symmetries and hence are called time reversal invariant momenta (TRIM) points.

2.5.3 Particle–hole symmetry

A condensed matter system is said to be particle–hole symmetric when the corresponding Hamiltonian commutes or anticommutes with the particle–hole symmetry operation. The operation is antilinear and hence differs from its close cousin charge conjugation, which is linear.[6] In a conventional sense, the superconductors possess particle–hole symmetry, and thus they are mostly involved in this context.

---

[6]There is a subtle point that requires a mention here. The charge conjugation is indeed antilinear in the first quantized theory, such as the Dirac equation in relativistic quantum mechanics. However, it is linear in the second quantized theory.

Let us use $\mathcal{C}$ for the particle–hole symmetry operation. It changes the creation operator $c^\dagger$ to annihilation operator c. ($\mathcal{C}$ stands for the charge conjugation symmetry which is often synonymously used with particle–hole symmetry.) Thus, it is precluded to be a symmetry operation for the purely free fermionic systems. For any self-adjoint single-particle Hamiltonian,

$$\mathcal{C}\mathcal{H}\mathcal{C}^{-1} = -\mathcal{H}.$$

The symmetry operation changes the electrons into holes and thus can be represented by

$$\mathcal{C} = \sigma_x \mathcal{K}.$$

Thus, it is anti-unitary in the sense $\mathcal{C}i\mathcal{C}^{-1} = -i$.

As an example, consider a generic Hamiltonian for a spinless superconductor in real space,

$$\mathcal{H} = \sum_{\langle ij \rangle}(t_{ij}c_i^\dagger c_j + h.c.) + \sum_{\langle ij \rangle}(\Delta c_i^\dagger c_j^\dagger + (\Delta^* c_j c_i), \tag{2.5.3.1}$$

where the first and the second terms denote kinetic and pairing terms, respectively. Reminding ourselves that the c operators anticommute, that is, $\{c_i^\dagger, c_j\} = \delta_{ij}$ and $\{c_i c_j\} = 0$. Clearly, $\mathcal{H}$ does not conserve the number of particles. But it does conserve one thing which is the parity of the number of electrons, namely whether the number of electrons is even or odd. In the particle–hole space spanned by all the creation and the annihilation operators (they are equal in number), namely

$$|\chi\rangle = |c_1 \ldots c_n, c_1^\dagger \ldots c_n^\dagger\rangle,$$

$\mathcal{H}$ can be expressed as

$$\mathcal{H} = \langle \chi | \mathcal{H}_{BdG} | \chi \rangle.$$

$\mathcal{H}_{BdG}$ is the Bogoliubov-de Gennes Hamiltonian, which can be written in the particle–hole basis as

$$\mathcal{H}_{BdG} = \begin{pmatrix} t & \Delta \\ -\Delta^* & -t^* \end{pmatrix}. \tag{2.5.3.2}$$

Here the creation operators are attributed to particles, and the annihilation operators denote holes, thereby doubling the number of degrees of freedom. As earlier,

$$\mathcal{C}\mathcal{H}_{BdG}\mathcal{C}^{-1} = -\mathcal{H}_{BdG}.$$

Thus, the symmetry operation anticommutes with the Hamiltonian. Also the energy spectrum corresponding to $\mathcal{H}_{BdG}$ is symmetric about the zero energy (Fermi level), yielding equal number of states for particles and holes. We shall come back to this when we discuss Kitaev chain in Chapter 3.

2.5.4   Chiral symmetry

In a usual sense, chiral symmetry refers to the sublattice symmetry in condensed matter physics. Thus, a natural realization of chiral symmetry is a system composed of two sublattices, where the sites of one sublattice only couple to the sites of the other sublattice. For the same reason, it is central to the discussion of symmetries in graphene (with nearest neighbour hopping)[7] owing to a bipartite structure, with each sublattice site containing a C atom. Thus, the chiral symmetry operation transforms a C atom from 'A' sublattice to a C atom in 'B' sublattice. However, in hexagonal boron nitride (h-BN), the chiral symmetry is broken owing to the presence of two different atoms, namely boron and nitrogen in 'A' and 'B' sites.

Further, in graphene there are two Dirac points where the low-energy dispersion is that of massless Dirac electrons (discussed elaborately in Chapter 4) that have opposite chirality in the sense that if an electron traverses one Dirac point in clockwise direction, it will do so anticlockwise for the other Dirac point. Thus, a chiral operation of the Hamiltonian for graphene switches one Dirac point to another.

Mathematically, the chiral symmetry for a system described by a Hamiltonian $\mathcal{H}$ is denoted as

$$\Gamma \mathcal{H} \Gamma^{-1} = -\mathcal{H},$$

where Γ denotes a unitary operation such that $\Gamma\Gamma^{\dagger} = \mathbb{1}$ and $\Gamma^2 = \mathbb{1}$. Γ does not involve spatial coordinates, and hence can be satisfied by crystalline systems with a boundary.

---

[7]A second neighbour hopping breaks the chiral symmetry.

3

Topology in One-Dimensional (1D) and Quasi-1D Models

In this chapter, we shall discuss three paradigmatic models that show symmetry-protected topological features and are resilient to local perturbations as long as the relevant symmetries are not disturbed. They are Su–Schrieffer–Heeger (SSH) model and a Kitaev chain with superconducting correlations in one-dimensional (1D) and a ladder system, known as the Creutz ladder in a quasi-1D setup.

3.1 Su–Schrieffer–Heeger (SSH) Model

3.1.1 Introduction

To make our concepts clear on the topological phase, and whether a model involves a topological phase transition, we apply it to the simplest model available in the literature. The SSH model denotes a paradigmatic 1D model that hosts a topological phase. It also possesses a physical realization in polyacetylene, which is a long chain organic polymer (polymerization of acetylene) with a formula $[C_2H_2]_n$ (shown in Fig. 3.1). The C–C bond lengths are measured by NMR spectroscopy technique and are found to be 1.36 Å and 1.44 Å for the double and the single bonds respectively. The chain consists of a number of methyne ($= CH-$) groups covalently bonded to yield a 1D structure, with each C-atom having a π electron. This renders connectivity to the polymer chain.

Possibly intrigued by this bond-length asymmetry, one can write down a tight-binding Hamiltonian of such a system with two different hopping parameters

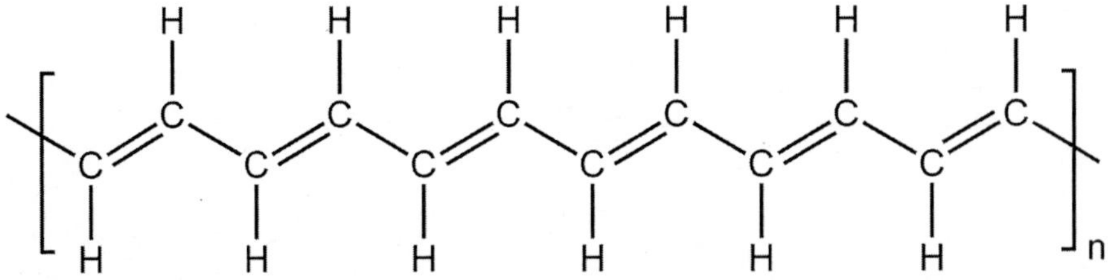

FIGURE 3.1. A polyacetylene chain with formula $(C_2H_2)_n$ is shown.

for spinless fermions hopping along the single and the double bonds. These *staggered* hopping amplitudes are represented by t_1 and t_2. Let us consider that the chain consists of N unit cells with two sites (that is, two C atoms) per unit cell and denote these two sites as A and B. The hopping between A and B sites in a cell be denoted by t_1, while those from B to A across the cell can be denoted by t_2. Because of the presence of a single π electron at each of the C atoms, the interparticle interaction effects are completely neglected. We shall show that the staggered hopping or the dimerization has got serious consequences for the topological properties of even such a simple model.

3.1.2 The SSH Hamiltonian

The above considerations yield the following Hamiltonian:

$$\mathcal{H} = -t_1 \sum_{n=1}^{N} (c_{n,A}^\dagger c_{n,B} + \text{h.c.}) - t_2 \sum_{n=1}^{N-1} (c_{n,B}^\dagger c_{n+1,A} + \text{h.c.}). \tag{3.1.2.1}$$

For simplicity and concreteness, t_1 and t_2 are assumed to be real and non-negative and $c_{n,\alpha}^\dagger (c_{n,\alpha})$ denote the electron creation (annihilation) operator at site n belonging to the α sublattice ($\alpha \in$ A, B).

It is clear that N denotes the total number of cells, which implies $M = 2N$, where M represents the total number of sites. Thus, for an open chain with M atoms, we have $t_M = 0$. In the site basis, the Hamiltonian can be explicitly written as

$$\mathcal{H} = (c_1^\dagger, c_2^\dagger, ..., c_M^\dagger) \begin{pmatrix} 0 & t_1 & 0 & . & . & 0 \\ t_1^* & 0 & t_2 & . & . & . \\ 0 & t_2^* & 0 & . & . & . \\ . & . & . & . & . & . \\ . & . & . & . & . & . \\ 0 & . & . & . & . & 0 \end{pmatrix} \begin{pmatrix} c_1 \\ c_2 \\ . \\ . \\ c_M \end{pmatrix} \tag{3.1.2.2}$$

If M is an even number, then $t_{M-1} = t_1$; otherwise, $t_{M-1} = t_2$.

We shall show in the following discussion that a staggered hopping is responsible for the opening of a gap in the dispersion, and subject to the fulfilment of a particular condition, the nature of the gap can be topological. To see that, let us study the band structure. We can Fourier transform the electron operators using

$$c_\alpha(k) = \sum_n e^{ikn} c_{n\alpha} \quad (\alpha \in A, B). \tag{3.1.2.3}$$

This yields a tight-binding Hamiltonian in the sublattice basis, namely (c_{kA}, c_{kB}) as

$$\mathcal{H} = \sum_k c_{k\alpha}^\dagger h_{\alpha\beta}(k) c_{k\beta}, \tag{3.1.2.4}$$

where

$$h_{\alpha\beta}(k) = \begin{pmatrix} 0 & t_1 + t_2 e^{-ik} \\ t_1 + t_2 e^{ik} & 0 \end{pmatrix} = \begin{pmatrix} 0 & f(k) \\ f^*(k) & 0 \end{pmatrix}. \tag{3.1.2.5}$$

The 2×2 structure of the matrix $h_{\alpha\beta}(k)$ allows us to write

$$h_{\alpha\beta}(k) = \mathbf{d}(k) \cdot \sigma, \tag{3.1.2.6}$$

where $\mathbf{d}(k)$ is a vector given by

$$\mathbf{d}(k) = (d_x(k), d_y(k), d_z(k)) = (t_1 + t_2 \cos k, t_2 \sin k, 0) \tag{3.1.2.7}$$

and $\sigma = (\sigma_x, \sigma_y, \sigma_z)$ denote the Pauli matrices. The energy dispersion is given by

$$E(k) = \pm|\mathbf{d}(k)| = \pm\sqrt{(t_1 + t_2 \cos k)^2 + t_2^2 \sin^2 k}. \tag{3.1.2.8}$$

A little manipulation of the terms inside the square root yields

$$E(k) = \pm\sqrt{(t_1 - t_2)^2 + 4t_1 t_2 \cos^2 \frac{k}{2}}, \tag{3.1.2.9}$$

where k is contained in the Brillouin zone (BZ), that is, $-\pi \leq k \leq +\pi$. The corresponding normalized eigenvectors are given by

$$|\psi_\pm\rangle = \frac{1}{\sqrt{2}} \begin{pmatrix} \pm e^{-i\phi(k)} \\ 1 \end{pmatrix}, \tag{3.1.2.10}$$

where

$$\phi(k) = tan^{-1}\left(\frac{t_2 \sin k}{t_1 + t_2 \cos k}\right).$$

We shall explore a few representative cases to make our ongoing discussion clear, namely

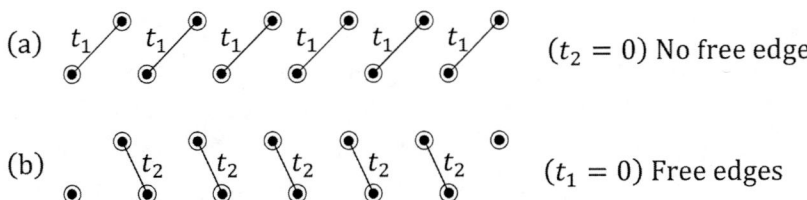

FIGURE 3.2. The extreme dimerized limit for the trivial (upper panel) and the topological (lower panel) phases of the SSH model are shown.

(i) $t_2 = 0$: Extreme dimerized limit (see the upper panel (a) of Fig. 3.2).
(ii) $t_1 > t_2$: Intra-cell hopping is larger than the inter-cell hopping.
(iii) $t_1 = t_2$: Intra-cell hopping is same as the inter-cell hopping.
(iv) $t_1 < t_2$: Intra-cell hopping is smaller than the inter-cell hopping.
(v) $t_1 = 0$: Extreme dimerized limit (however, different than (i) (see the lower panel (b) of Fig. 3.2)).

We plot the band structure and the components of the **d**-vector as a function of the crystal momentum k. The purpose is to define a bulk winding number, which is the topological invariant here. The plot d_x vs d_y for k in the BZ defines a surface (except for the critical case, $t_1 = t_2$). The topological properties of the system will be decided based on whether the surface encloses the origin. Further, the unit vector $\hat{\mathbf{d}}$ defines the direction of the d vector via $\hat{\mathbf{d}} = \mathbf{d}/|\mathbf{d}|$. At half filling, the lower band is filled. The two bands are gapped by an amount $2\delta t$, where $\delta t = |t_1 - t_2|$ at $k = \pm\pi$. This is also an insulating phase. However, we shall see that this phase is distinct from the case (ii).

3.1.3 Topological properties

The SSH chain hosts both bulk and edge states. The distinction between the bulk and the edge can be understood from the real space analysis. The plot for the energy spectrum as a function of the ratio t_1/t_2 (see Fig. 3.3) shows that the zero modes start to appear just beyond the critical point ($t_1 = t_2$). Prior to that, for $0 \le t_2/t_1 \le 1$, the system behaves like a trivial insulator with a bulk band gap. The gap closes at $t_1 = t_2$, and eventually for $t_2 > t_1$, the bulk gap opens again; however, a pair of zero modes appear in the spectrum. These zero modes yield a topological character to the phase. They originate from the two solitary C atoms that reside at the two edges of the chain. The fact that these zero modes indeed arise out of the edges are shown in Fig. 3.4 via plotting the probability densities $|\psi_i|^2$ at all sites of the chain. The amplitudes at the left and the right edges are characterized by finite values, while they vanish everywhere in the bulk of the chain. On

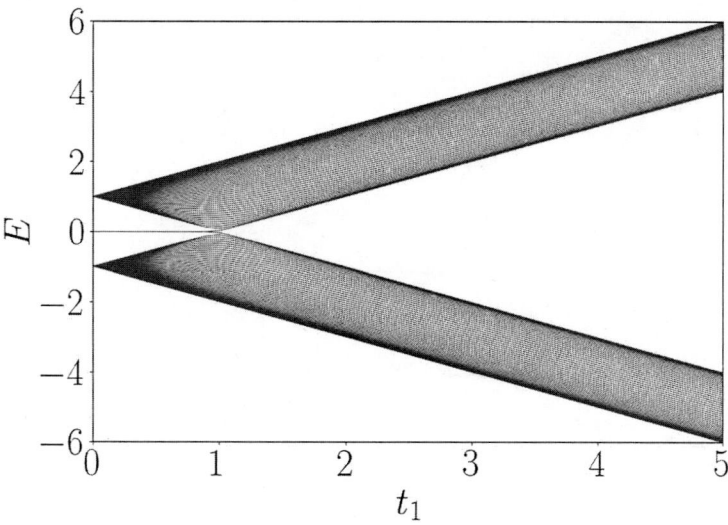

FIGURE 3.3. The energy is plotted as a function of t_1 (in unit of t_2). A zero mode exists for $t_1 \leq 1$ as can be clearly noted. The presence of zero implies finite values of IPR (see text).

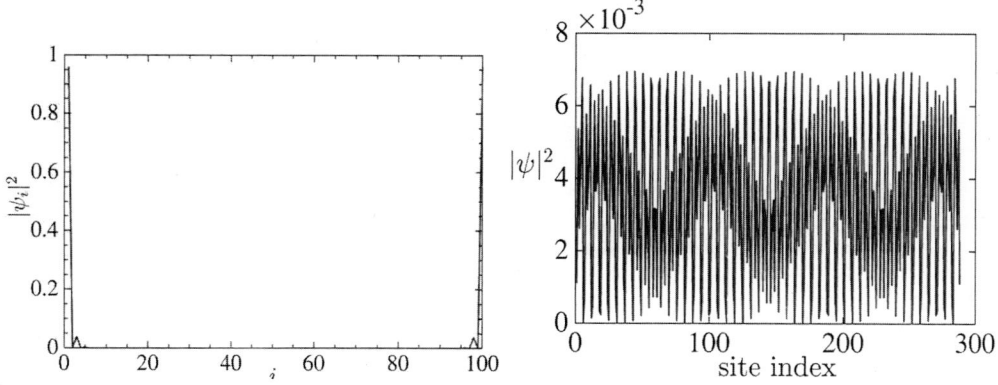

FIGURE 3.4. The probability amplitude $|\psi|^2$ is plotted as a function of sites of the chain. The left-hand panel shows localization of the zero-energy edge modes for $t_2 > t_1$, while the right-hand panel shows the extended bulk modes of the system. Here we have taken the length, $L = 100$.

the other hand, the extended bulk modes are shown in the right-hand panel of Fig. 3.4, emphasizing the topological nature of the system. Further, to emphasize on the robustness of the edge modes, we show the inverse participation ratio (IPR)

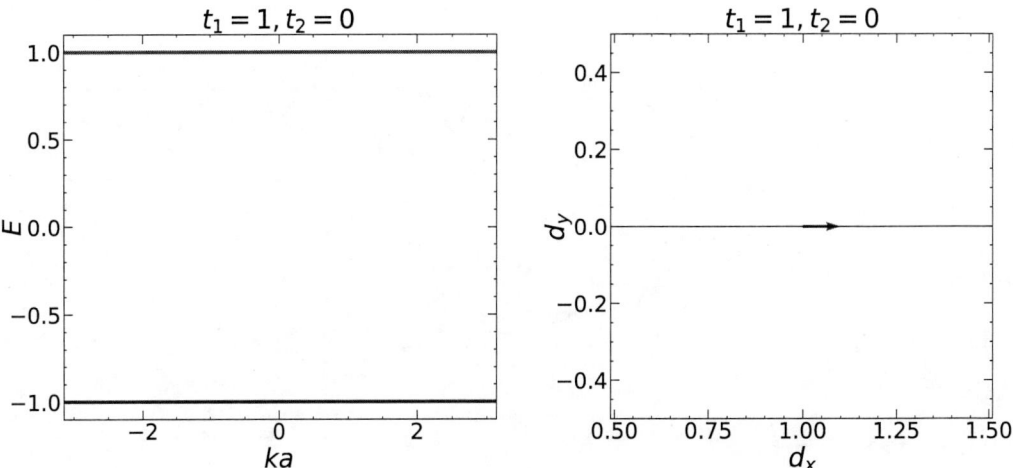

FIGURE 3.5. The band structure and d-vectors are plotted corresponding to $t_1 = 1$, $t_2 = 0$.

defined by

$$\text{IPR} = \sum_{i=1}^{L} |\psi_i|^4, \qquad (3.1.3.1)$$

IPR $= 0$ or 1 denotes the extended or the localized phases. However, these extreme values (namely 0 and 1) can only be obtained in the thermodynamic limit ($L \to \infty$). Here we show the edge modes at positive and negative energies in Fig. 3.5, which are characterized by finite values of IPR. Evidently, the zero modes are seen to be localized.

Let us return to the behaviour of the vector $\mathbf{d}(k)$. The components $((t_1 + t_2 \cos k), t_2 \sin k, 0)$ in the BZ defined by $-\pi \leq k \leq +\pi$ of $\mathbf{d}(k)$ denote the eigenstates with the energy spectrum given by

$$E(k) = |\mathbf{d}(k)|.$$

Corresponding to one of the cases, namely $t_2 > t_1$, the vector $\mathbf{d}(k)$ winds about the origin, while for the other, $t_2 < t_1$, it does not. The origin of the d_x-d_y plane is $\mathbf{d}(k) = 0$ and denotes the gapless (critical) condition. Based on the above information, it is possible to define a winding number, ν, which would tell us whether the trajectory of $\mathbf{d}(k)$ winds the origin, as k is varied over the BZ. Thus, the winding number is capable of distinguishing the two seemingly equivalent (gapped) scenarios.

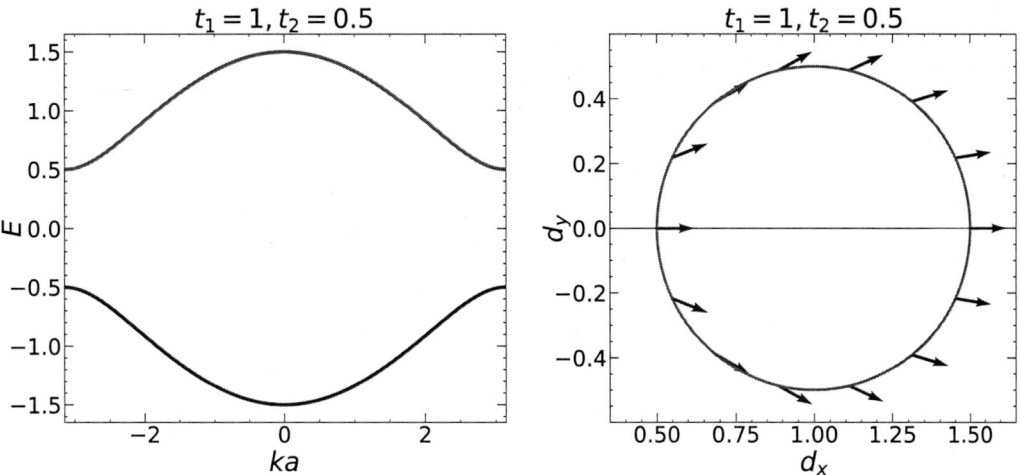

FIGURE 3.6. The band structure and d-vectors are plotted corresponding to $t_1 = 1$, $t_2 = 0.5$.

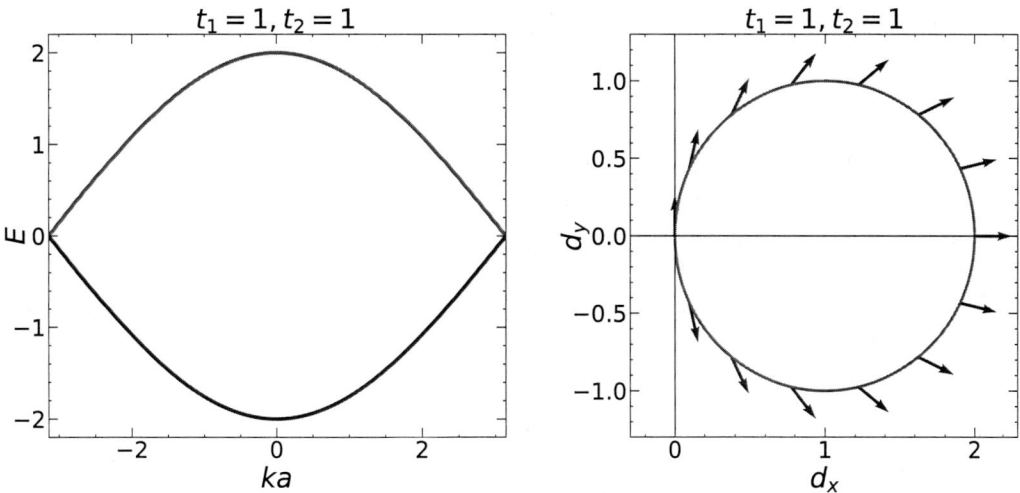

FIGURE 3.7. The band structure and d-vectors are plotted corresponding to $t_1 = 1$, $t_2 = 1$, which is a simple tight-binding chain.

Mathematically, the winding number, ν, can be written down using the unit $\hat{\mathbf{d}}$ vector defined via

$$\hat{\mathbf{d}} = \frac{\mathbf{d}(k)}{|\mathbf{d}(k)|}. \tag{3.1.3.2}$$

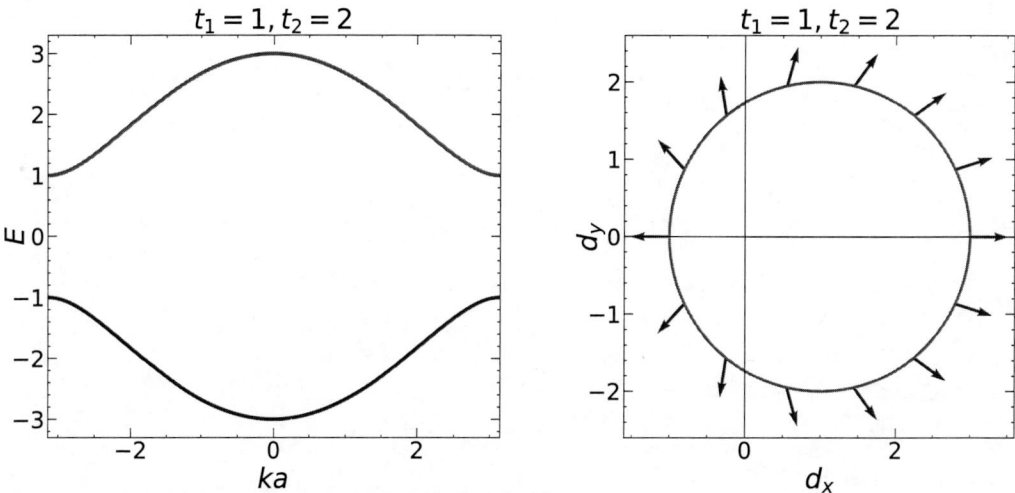

FIGURE 3.8. The band structure and d-vectors are plotted corresponding to $t_1 = 1$, $t_2 = 2$.

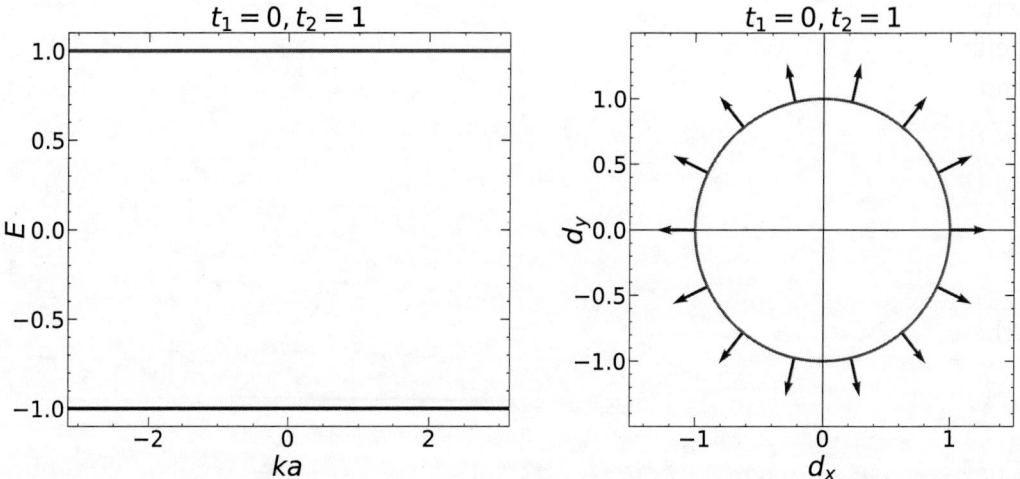

FIGURE 3.9. The band structure and d-vectors are plotted corresponding to $t_1 = 0$, $t_2 = 1$.

One can now define ν using

$$\nu = \frac{1}{2\pi} \int_{-\pi}^{+\pi} \left(\hat{\mathbf{d}} \times \frac{d}{dk} \hat{\mathbf{d}} \right)_z dk. \qquad (3.1.3.3)$$

Let us justify how the above expression on the RHS denotes the winding number. Writing it more explicitly,

$$\nu = \frac{1}{2\pi} \int_{-\pi}^{\pi} \frac{\mathbf{d}(k)}{|\mathbf{d}(k)|} \times \frac{d}{dk} \frac{\mathbf{d}(k)}{|\mathbf{d}(k)|} dk$$

$$= \frac{1}{2\pi} \int_{-\pi}^{\pi} \frac{\mathbf{d}(k)}{|\mathbf{d}(k)|} \times \left(\frac{d}{dk} \frac{\mathbf{d}(k)}{|\mathbf{d}(k)|} - \frac{\mathbf{d}(k) \frac{d}{dk} |\mathbf{d}(k)|}{|\mathbf{d}(k)|^2} \right) dk,$$

since $\mathbf{d}(k) \times \mathbf{d}(k) = 0$

$$\nu = \frac{1}{2\pi} \int_{-\pi}^{\pi} \frac{\mathbf{d} \times \frac{d}{dk}\mathbf{d}(k)}{|\mathbf{d}(k)|^2} dk$$

$$= \frac{1}{2\pi} \int_{-\pi}^{\pi} \frac{\mathbf{d}(k) \times \delta\mathbf{d}(k)}{|\mathbf{d}(\mathbf{k})|^2} dk$$

$$\frac{d}{dk}\mathbf{d}(k) = \frac{d}{dk}(\mathbf{d}_0 + k\delta\mathbf{d}),$$

where in the last line we have used a Taylor expansion of $\frac{d}{dk}\mathbf{d}(k)$. From the definition of the cross-product, $\mathbf{d}(k) \times \delta\mathbf{d}(k)$ is the angle (in radian) between $\mathbf{d}(k)$ and $\mathbf{d} + \delta\mathbf{d}$. Thus, integrating this over the BZ yields 2π, which when divided by 2π gives 1.

Another useful form for the winding number is given by

$$\nu = \frac{1}{2\pi i} \int_{-\pi}^{+\pi} dk \frac{d}{dk} \log f(k), \tag{3.1.3.4}$$

where $f(k) = t_1 + t_2 e^{-ik}$. Thus,

$$\log f(k) = \log(|f|) e^{i \arg(f)}.$$

Consequently, the winding number becomes

$$\nu = \frac{1}{2\pi} \arg(f)\big|_{-\pi}^{+\pi} \tag{3.1.3.5}$$

$$= 1 \text{ or } 0$$

depending on whether $\arg(f)$ falls in the region of integration, that is, it encloses the origin. The winding number is 1 for the topological phase and 0 for the trivial phase.

We correlate the band structure with the corresponding winding number calculated for the various cases mentioned above. In the extreme dimerized limit, namely $t_2 = 0$ and $t_1 = 1$, we get two flat bands at $E = \pm 1$, and the **d**-vector is simply shown by an arrow in Fig. 3.5. For $t_1 > t_2$, the spectrum is gapped and corresponds to a trivial insulator because of the absence of winding of the **d**-vector as shown in Fig. 3.6. Further, the undimerized tight-binding chain ($t_1 = t_2$) is shown in Fig. 3.7, where the gap closes and the tip of the **d**-vector that executes a circle in the d_x–d_y plane just touches the origin at the left but does not wind it. The fourth case, namely $t_1 < t_2$, again shows a spectral gap, but it is topological in nature as shown by the winding of the **d**-vector in Fig. 3.8. Finally, the other dimerized case, that is, $t_1 = 0$ and $t_2 = 1$ (see Fig. 3.9), again shows two flat bands at $E = \pm 1$; however, the trajectory of the d-vector shown by the circle in the right-hand panel of Fig. 3.9 winds the origin and hence denotes a topological scenario.

One can also define a Zak phase Φ_Z [10] (another geometric phase, similar to the Berry phase) defined via[1]

$$\Phi_Z = i \oint \langle \psi | \nabla_k | \psi \rangle \, dk. \tag{3.1.3.6}$$

Using

$$|\psi_\pm\rangle = \frac{1}{\sqrt{2}} \begin{pmatrix} \pm e^{-i\phi(k)} \\ 1 \end{pmatrix} \tag{3.1.3.7}$$

$$\phi_Z = \frac{1}{2} \oint \frac{d}{dk}\phi(k)\,dk = \pm\pi \text{ or } 0, \tag{3.1.3.8}$$

which are the values respectively for $t_2 > t_1$ and $t_2 < t_1$.

Please note that we have obtained this result without plugging in the explicit form of $\phi(k)$, since the result should be independent of the form of $\phi(k)$. However, if we consider the explicit form of $\phi(k)$, namely

$$\phi(k) = tan^{-1} \left(\frac{t_2 \sin k}{t_1 + t_2 \cos k} \right),$$

it throws some subtlety that we need to take care of. If we are in the trivial phase, then the inverse tangent function is present in the first and fourth quadrants because of $t_1 > t_2$. Here the function does not acquire any extra factor because of

---

[1]Usually geometric phases that characterize the topological properties of the band structure play a crucial role in the band theory of solids. See Ref. [10].

which $\phi_Z = 0$. For the topological phase ($t_2 > t_1$), that is, when the inverse tangent function is in the second quadrant, it picks up a phase $\pi - \tan^{-1} x$, while in the third quadrant, the corresponding value is $\pi + \tan^{-1} x$. This can be seen from the sharp change of $\phi(k)$ twice in BZ as seen in Fig. 3.10. This means that the inverse tangent function acquires an extra phase of 2π. This yields $\phi_Z = -\pi$. Frankly, the negative sign does not mean anything specific. It arises because we have chosen a positive sign for the wavefunction. Here the winding number and the Zak phase are related via $\nu = -\phi_Z/\pi$. Think about $\mathbf{d}(k)$ and ϕ_Z, both of which are obtained from the bulk of the material, yet giving information about the edges of the system. This is traditionally referred to as the *bulk–boundary correspondence* (we shall discuss more about it later). It may also be noted that the behaviour of $\phi(k)$ is smooth over the BZ corresponding to the trivial case (see Fig. 3.11).

From the preceding discussion, it is clear that the two apparently similar insulating phases are topologically different. We have shown that the winding numbers are different (finite in the topological phase and zero in the trivial one) and so are the Zak phases. However, what does it physically entail having different values for the winding number or the Zak phase? Suppose we smoothly deform the Hamiltonian corresponding to the SSH chain via tuning the hopping parameters, t_1 and t_2, we shall get two insulating phases for $t_2 > t_1$ and $t_2 < t_1$ all the while keeping the symmetries preserved, and the band gap around $E = 0$ is finite. In tuning from one limit to the other, we have to cross the origin in the d_x–d_y plane (note that $d_z \equiv 0$). This implies that at the intermediate stage, one would obtain the eigensolutions corresponding to the trajectory of $\mathbf{d}(k)$ in the d_x–d_y plane through the origin. Thus, a smooth transition from one insulating phase to another is impossible without closing the gap, or satisfying the metallic condition $t_1 = t_2$. Hence, quite apparently, there is a topological phase transition occurring here.

It is clear that the above discussion will be invalid if there is a z-component of the $\mathbf{d}(k)$ (or a term proportional to σ_z) present in the system. A simple way to incorporate such a term is through inclusion of an onsite potential. The onsite potential will destroy the zero modes, thereby making it meaningless to talk about the topological properties of the system. However, a disorder in the off-diagonal (hopping) term would retain the zero modes, and hence the system should have a transition from a topological to a trivial phase. The reason for such a distinction between the diagonal and the off-diagonal terms arises because of certain fundamental symmetries that the system possesses. We shall discuss them below.

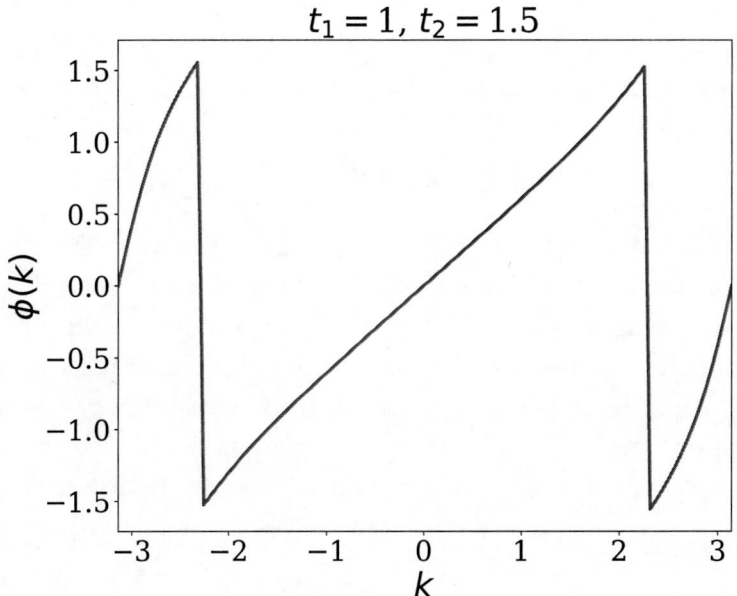

FIGURE 3.10. ϕ is plotted as a function of k for the topological phase. There are abrupt jumps in the behaviour of ϕ.

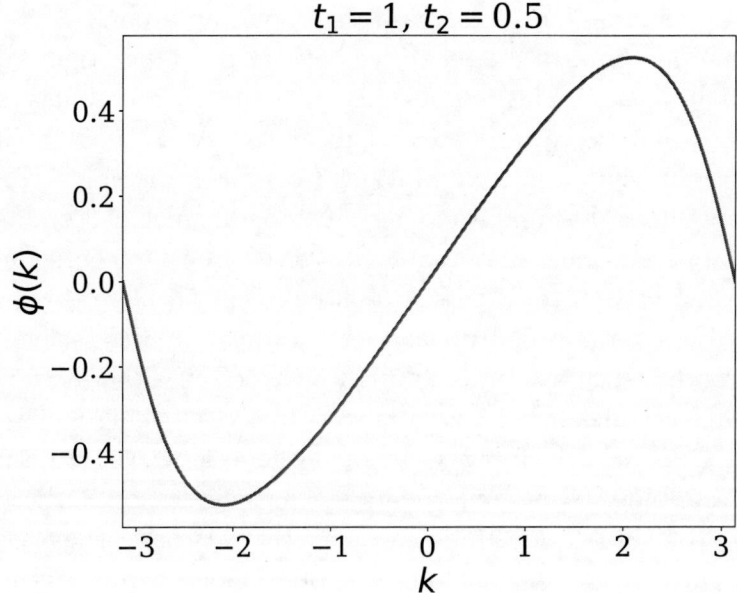

FIGURE 3.11. ϕ is plotted as a function of k for the trivial phase. It smoothly varies from $-\pi$ to $+\pi$.

3.1.4 Chiral symmetry of the SSH model

In standard quantum mechanics, the symmetry of a Hamiltonian $\mathcal{H}$ is represented by

$$U\mathcal{H}U^\dagger = \mathcal{H}, \text{ or } U\mathcal{H} = \mathcal{H}U, \text{ or } [\mathcal{H}, U] = 0, \tag{3.1.4.1}$$

where U denotes a unitary operator. This implies that U and $\mathcal{H}$ have the same eigenstates and hence can be diagonalized simultaneously. However, in general for topological insulators, such usual unitary symmetries do not have interesting consequences. The reason being it is mostly possible to make the Hamiltonian block diagonal, thereby reducing the problem to be confined in a single block. In the case of massless Dirac problems, one usually runs out of unitary symmetries and is left with an irreducible block Hamiltonian, which cannot be diagonalized. Thus, for the SSH model, there is a different symmetry, which is called the chiral symmetry that is operative here.[2] Here $\mathcal{H}$ obeys

$$\Gamma\mathcal{H}\Gamma^\dagger = -\mathcal{H}, \text{ or } \Gamma\mathcal{H} = -\mathcal{H}\Gamma, \text{ or } \{\mathcal{H}, \Gamma\} = 0. \tag{3.1.4.2}$$

Here Γ is a unitary operator corresponding to the chiral symmetry. Instead of commuting, it anticommutes with the Hamiltonian. Further, Γ is unitary and Hermitian, implying

$$\Gamma = \Gamma^\dagger \text{ or } \Gamma^\dagger\Gamma = \Gamma^2 = \mathbf{1}, \tag{3.1.4.3}$$

where $\mathbf{1}$ denotes an identity matrix. The above requirement raises a possibility that $\Gamma = e^{i\phi}$, where ϕ is an arbitrary phase. However, this possibility can be eliminated by redefining $\Gamma \to \Gamma e^{-i\phi/2}$. A second requirement is that Γ is a local operator. Thus, the matrix elements of Γ survive only within each unit cell, and between the cells they vanish. Hence, for the SSH model, the chiral symmetry is equivalent to the sublattice symmetry, which can be expressed through the projectors P_A and P_B corresponding to the A and B sublattices, namely

$$P_A = \frac{1}{2}(\mathbf{1} + \Gamma); \quad P_B = \frac{1}{2}(\mathbf{1} - \Gamma). \tag{3.1.4.4}$$

It can be checked that

$$P_A + P_B = \mathbf{1} \text{ and } P_A \cdot P_B = 0.$$

---

[2]In condensed matter physics, bipartite systems with nearest neighbour hopping, that is, when hopping connects sites with opposite sublattices, obey chiral symmetry.

It is also possible to show that

$$P_A \mathcal{H} P_A = P_B \mathcal{H} P_B = 0. \tag{3.1.4.5}$$

The consequence of the chiral symmetry results in a symmetric energy spectrum. That is, corresponding to an energy E, there is a chiral partner with energy $-E$. This fact can be seen from the following:

$$\mathcal{H}|\psi\rangle = E|\psi\rangle \tag{3.1.4.6}$$
$$\mathcal{H}\Gamma|\psi\rangle = -\Gamma\mathcal{H}|\psi\rangle = -\Gamma E|\psi\rangle = -E\Gamma|\psi\rangle.$$

Of course, the above argument is true for $E \neq 0$. Since the SSH model hosts zero modes, one of them is a partner of the other. Besides, for all $E \neq 0$, $|\psi\rangle$ and $\Gamma|\psi\rangle$ correspond to distinct and orthogonal eigenstates, which suggests that every non-zero eigenstate of $\mathcal{H}$ derives equal contribution from both the sublattices, that is,

$$\langle\psi|\Gamma|\psi\rangle = \langle\psi|P_A|\psi\rangle - \langle\psi|P_B|\psi\rangle = 0. \tag{3.1.4.7}$$

Whereas, for $E = 0$, $\mathcal{H}|\psi\rangle = 0$. Thus,

$$\mathcal{H}P_{A/B}|\psi\rangle = \mathcal{H}\left[|\psi\rangle + \Gamma|\psi\rangle\right] = 0. \tag{3.1.4.8}$$

Thus, the zero energy eigenstates are eigenstates of Γ and hence are chiral symmetric partners of themselves. It is also owing to the robustness of the chiral symmetry that the zero modes are robust.

One can also define operators for the sublattice symmetry, namely

$$\Sigma_z = P_A - P_B. \tag{3.1.4.9}$$

It can be shown that $\Sigma_z \mathcal{H} \Sigma_z = -\mathcal{H}$, which is similar to the symmetry relation stated earlier, $\Gamma\mathcal{H}\Gamma = -\mathcal{H}$. Thus, the chiral symmetry of the SSH model is a re-statement of the sublattice symmetry of the Hamiltonian.

In simple language, the chiral symmetry operator Γ is actually the z-component of the Pauli matrix σ_z, which yields

$$\sigma_z \mathcal{H} \sigma_z = -\mathcal{H}. \tag{3.1.4.10}$$

A direct multiplication of the three matrices on the LHS can be performed for the proof. We have presented above a crisp description of the SSH model, which even being simple enough, possesses both trivial and topological phases where the latter shows up via the presence of robust[3] zero energy edge modes, along with a finite value of the winding number. Further, the band structure shows that a phase transition occurs from a trivial to a topological phase (or vice versa) through a gap closing point, where the staggered hopping amplitudes are equal ($t_1 = t_2$).

3.2 Kitaev model

3.2.1 Introduction

In a 1D dimerized chain comprising of two atoms per unit cell within a tight-binding approximation, which is known as the SSH model, there are localized zero energy modes. These modes are robust against adiabatic deformation of the Hamiltonian. Thus, inclusion of disorder (or defect) that respects the chiral symmetry does not harm these zero modes. However, it is difficult to satisfy chiral symmetry in real physical systems.

In an analogous model, however, with additional ingredients, the so-called Kitaev model in 1D the symmetry that protects the topological features is much more physical than the chiral symmetry. In fact, it is the particle–hole symmetry that plays a crucial role here, and is also inherent to the mean field description of superconductors. Thus, the Kitaev chain is a more realistic model to access the robust zero energy edge modes, the so-called Majorana zero modes. A brief description will be included at the end of this chapter to discuss their physical realizability in experiments, and their possible applications in using the degenerate quantum states at zero energy for storing quantum information.

The system comprises a 1D p-wave superconductor introduced by Kitaev[11], where the superconducting correlations occur between spinless (or spin-polarized) fermions at neighbouring sites in the chain, as opposed to the onsite pairing discussed in the context of the more familiar BCS (s-wave) superconductors. In this model, the Majorana fermions (MFs), which arise as real solutions to the Dirac equation and are their own antiparticles, emerge in a simple and intuitive fashion. The search for Majorana particles remained inconclusive as they have never been observed in nature. Thus, the Kitaev model also serves as a platform for realizing Majorana particles in condensed matter systems. In the following discussion, let us begin with the description of a two-site system, which essentially contains all the information that we need.

---

[3]The edge modes are robust as long as the chiral symmetry is intact.

3.2.2 Two-site Kitaev chain

In this section, we shall study the edge and the bulk properties of a two-site Kitaev chain with a view to explore the topological properties of the excitation spectrum. At each site, there is a single spinless fermion that is coupled to a p-wave superconductor. It may be noted that by virtue of the fermions being spinless (or spin polarized), a conventional s-wave pairing for fermions belonging to the same site is not possible. The Hamiltonian for such a system is written as

$$\mathcal{H}_{2site} = -\mu c_1^\dagger c_1 - \mu c_2^\dagger c_2 - t(c_1^\dagger c_2 + \text{h.c.}) + \Delta(c_1^\dagger c_2^\dagger + \text{h.c.}), \qquad (3.2.2.1)$$

where μ denotes the chemical potential, t is the hopping term among the neighbouring sites and Δ is the p-wave superconducting order parameter. We assume all of these energy scales are real and positive. In fact, Δ is usually a complex quantity with an amplitude and a phase. However, taking it either real or complex does not have much consequence on our discussion.

The Hamiltonian can be written in the basis spanned by $\{c_1^\dagger, c_1, c_2^\dagger, c_2\}$. Explicitly write it in this basis, we get

$$\mathcal{H}_{2site} = \begin{pmatrix} c_1^\dagger & c_1 & c_2^\dagger & c_2 \end{pmatrix} \begin{pmatrix} -\mu & 0 & -t & \Delta \\ 0 & \mu & -\Delta & t \\ -t & -\Delta & -\mu & 0 \\ \Delta & t & 0 & \mu \end{pmatrix} \begin{pmatrix} c_1 \\ c_1^\dagger \\ c_2 \\ c_2^\dagger \end{pmatrix}. \qquad (3.2.2.2)$$

In the Dirac notation, we can use the particle–hole representation at each site by writing $c_1^\dagger = |1e\rangle$, $c_1 = \langle 1h|$, $c_2^\dagger = |2e\rangle$ and $c_2 = \langle 2h|$, where 1 and 2 refer to the sites and e and h refer to the electron (particle) and the hole states respectively. The electron–hole description that is natural to a superconductor becomes apparent in this basis and the Hamiltonian can be written as

$$\begin{aligned} \mathcal{H}_{2site} = & - \mu(|1e\rangle\langle 1e| + |1h\rangle\langle 1h| + |2e\rangle\langle 2e| + |2h\rangle\langle 2h|) \qquad (3.2.2.3) \\ & - t(|1,e\rangle\langle 2e| - |1h\rangle\langle 2h| + \text{h.c.}) \\ & + \Delta(|1e\rangle\langle 2h| - |2e\rangle\langle 1h| + \text{h.c.}). \end{aligned}$$

Diagonalization of the above 4×4 Hamiltonian yields the eigensolutions. The eigenvalues are given by

$$E = (t + \sqrt{(\Delta^2 + \mu^2)}), \ (t - \sqrt{(\Delta^2 + \mu^2)}), \ -(t - \sqrt{(\Delta^2 + \mu^2)}), \ -(t + \sqrt{(\Delta^2 + \mu^2)}). \qquad (3.2.2.4)$$

The corresponding eigenvectors are

$$
\begin{pmatrix} \frac{(t+\sqrt{(\Delta^2+\mu^2)})}{\Delta} - \frac{(\mu+t)}{\Delta} \\ 1 \\ \frac{(\mu+t)}{\Delta}) - \frac{(t+\sqrt{(\Delta^2+\mu^2)})}{\Delta} \\ 1 \end{pmatrix}, \begin{pmatrix} \frac{(t-\sqrt{(\Delta^2+\mu^2)})}{\Delta} - \frac{(\mu+t)}{\Delta} \\ 1 \\ \frac{(\mu+t)}{\Delta}) - \frac{(t-\sqrt{(\Delta^2+\mu^2)})}{\Delta} \\ 1 \end{pmatrix},
$$

$$
\begin{pmatrix} \frac{(-t+\sqrt{(\Delta^2+\mu^2)})}{\Delta} - \frac{(\mu-t)}{\Delta} \\ -1 \\ \frac{(-t+\sqrt{(\Delta^2+\mu^2)})}{\Delta} - \frac{(\mu-t)}{\Delta}) \\ 1 \end{pmatrix}, \begin{pmatrix} \frac{(-t-\sqrt{(\Delta^2+\mu^2)})}{\Delta} - \frac{(\mu-t)}{\Delta} \\ -1 \\ \frac{(-t-\sqrt{(\Delta^2+\mu^2)})}{\Delta} - \frac{(\mu-t)}{\Delta}) \\ 1 \end{pmatrix}. \tag{3.2.2.5}
$$

Writing down the Hamiltonian in the form as it appears in Eq. (3.2.2.3) has the advantage that it can be trivially extended for a chain containing N sites and can be written as in the following:

$$
\mathcal{H} = \sum_n [-\mu \left(|n,e\rangle\langle n,e| + |n,h\rangle\langle n,h| \right) \tag{3.2.2.6}
$$
$$
- t \left(|n,e\rangle\langle n+1,e| - |n,h\rangle\langle n+1,h| + \text{h.c.} \right)
$$
$$
+ \Delta \left(|n,e\rangle\langle n+1,h| - |n+1,e\rangle\langle n,h| + \text{h.c.} \right)]
$$

$$
= \sum_n [-\mu \left(|n,e\rangle\langle n,e| + |n,h\rangle\langle n,h| \right) \tag{3.2.2.7}
$$
$$
+ (-t+\Delta) \left(|n,e\rangle\langle n+1,e| + \text{h.c.} \right)
$$
$$
+ (t-\Delta) \left(|n,h\rangle\langle n+1,h| + \text{h.c.} \right)].
$$

$$
\tag{3.2.2.8}
$$

The above form appears as a tight-binding Hamiltonian with two atoms per unit cell. A Fourier transform can be done using

$$
c_k = \sum_j e^{ikj} c_j; \quad c_k^\dagger = \sum_j e^{-ikj} c_j^\dagger, \tag{3.2.2.9}
$$

where k is the wave vector defined within the BZ $[-\pi : +\pi]$. Using the Fourier transformed operators and noting that

$$
-e^{-ik} c_k^\dagger c_{-k}^\dagger = +e^{-ik} c_k^\dagger c_{-k}^\dagger,
$$

one gets in the $\{c_k, c_k^\dagger\}$ basis

$$
\mathcal{H} = (-\mu - 2t \cos k)\sigma_z - (2\Delta \sin k \sigma_y) \tag{3.2.2.10}
$$
$$
= \mathbf{d}(k).\sigma,
$$

where σ denotes the components of the Pauli matrix and

$$\mathbf{d}(k) = (0, -2\Delta \sin k, -\mu - 2t \cos k).$$

The form is similar to a 2D massless Dirac Hamiltonian (similar to that of graphene). However, it needs to be noted that the components of σ do not represent spin degrees of freedom (recall that we are considering spinless fermions) and instead denote particle–hole degrees of freedom. It is now easy to diagonalize the 2×2 Hamiltonian whose eigensolutions are given by

$$E_k = \pm\sqrt{(-\mu - 2t \cos k)^2 + 4\Delta^2 \sin^2 k} \qquad (3.2.2.11)$$

with the eigenvectors as

$$\begin{pmatrix} u(k) \\ v(k) \end{pmatrix} = \begin{pmatrix} \frac{\sqrt{4t^2 \cos^2 k + 4\mu t \cos k + 4\Delta^2 \sin^2 k + i\mu^2}}{2\Delta \sin k} - \frac{(\mu + 2t \cos k)i}{2\Delta \sin k} \\ 1 \end{pmatrix};$$

$$\begin{pmatrix} \frac{-\sqrt{4t^2 \cos^2 k + 4\mu t \cos k + 4\Delta^2 \sin^2 k + i\mu^2}}{2\Delta \sin k} - \frac{(\mu + 2t \cos k)i}{2\Delta \sin k} \\ 1 \end{pmatrix}. \qquad (3.2.2.12)$$

It may be noted that the eigenvectors are in the standard format of a mean field solution of a superconductor.[4] Instead of solving the model for its topological characteristics, we shall discuss the symmetry properties first in the following section.

3.2.3   Particle–hole symmetry of the Kitaev model

Generally speaking, particle–hole symmetry implies for a state defined by energy and momentum (E, k), there is always a partner with $(-E, -k)$. This is reflected in the symmetry of the diagonal elements as seen in the Hamiltonian written in the k-space, namely in Eq. (3.2.2.10), where the cosine term is symmetric under the transformation of $k \to -k$. This is also an artefact of the p-wave superconducting term written in the mean field form which ensures particle–hole symmetry of the model.

Let us discuss the particle–hole symmetry of the Kitaev model in detail. The operator that tests the particle–hole symmetry, which we denote by $\Sigma = \sigma_x \mathcal{K}$, where σ_x is the x-component of the Pauli matrix (that acts on the particle–hole degrees of freedom) and $\mathcal{K}$ is the complex conjugation operator, such that

---

[4]These mean field solutions correspond to the Bogoliubov-de Gennes equations, which yield an equivalent description as the BCS theory for conventional superconductors.

$\mathcal{K}i\mathcal{K} = -i$. Σ is an anti-unitary operator that transforms the Hamiltonian in Eq. (3.2.2.10) as

$$\Sigma\mathcal{H}\Sigma = (\sigma_x\mathcal{K})\mathcal{H}(k)(\sigma_x\mathcal{K}) = -\mathcal{H}, \qquad (3.2.3.1)$$

where $(\sigma_x\mathcal{K})^2 = 1$. For a closer introspection, let us consider a wavefunction ψ at certain given values of an energy and momentum, (E, k). As per the discussion above, there will be a particle–hole partner ψ' with energy and momentum as $(-E, -k)$. Thus,

$$\psi' = \sigma_x\mathcal{K}\psi. \qquad (3.2.3.2)$$

We can write

$$(\sigma_x\mathcal{K})\mathcal{H}(k)\psi = E\sigma_x\mathcal{K}\psi. \qquad (3.2.3.3)$$

Inserting $(\sigma_x\mathcal{K})(\sigma_x\mathcal{K})$ $(= 1)$ between $\mathcal{H}(k)$ and ψ, one gets

$$(\sigma_x\mathcal{K})\mathcal{H}(k)(\sigma_x\mathcal{K})(\sigma_x\mathcal{K})\psi = E\sigma_x\mathcal{K}\psi. \qquad (3.2.3.4)$$

Substituting ψ' from Eq. (3.2.3.2),

$$(\sigma_x\mathcal{K})\mathcal{H}(k)(\sigma_x\mathcal{K})\psi' = E\psi'. \qquad (3.2.3.5)$$

Or,

$$\sigma_x\mathcal{H}^*(k)\sigma_x\psi' = E\psi'. \qquad (3.2.3.6)$$

Since $\sigma_x\mathcal{H}^*(k)\sigma_x = -\mathcal{H}(-k)$,

$$\mathcal{H}(-k)\psi' = -E\psi'. \qquad (3.2.3.7)$$

This proves the existence of a time-reversed partner under the particle–hole transformation.

It may be noted from above that corresponding to $E > 0$, the partners are orthogonal, that is, $\langle\psi|\psi'\rangle = 0$, since they have different eigenvalues. However, for $E = 0$, the situation is more interesting. ψ and ψ' are now identical. These are called the Majorana mode for which $\sigma_x\mathcal{K}\psi = \psi$.

For the sake of convenience and concreteness, we list out the parameters μ, t and Δ in Table 3.1 for discussing the topological and the trivial cases.

This brings us to the notion of topology in a Kitaev model that is familiar in systems where the boundary behaves differently compared to the bulk of the

TABLE 3.1. Different choices for the parameters μ, t and Δ for showing different phases of the Kitaev model.

μ	t	Δ
0	1	1
1	1	0
1	0	1
1	1	1
1	0.5	0
1	0.3	0
1	0.5	1
1	0	0

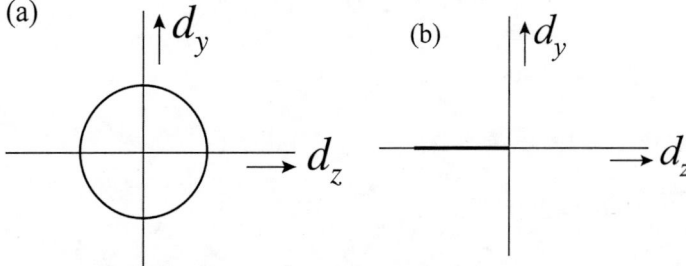

FIGURE 3.12. The *d*-vectors corresponding to the trivial and the topological phases are plotted.

system. This will become clear if we look at the $\mathbf{d}(k)$ vector defined in the yz-plane where the components are $d_y(k) = -2\Delta \sin k$ and $d_z(k) = -\mu - 2t \cos k$. The tip of the $\mathbf{d}$ vector may encircle the origin depending on the values of the parameter yielding an integer winding number.[5] There arise two situations governed by the value of d_z, namely (i) $d_z =$ positive, which results from $\mu > -2t$, and (ii) d_z to be negative for $\mu < -2t$. Representations from each of the cases may be taken as:
(a) $\mu = 0, t = \Delta = 1$ (a priori, this is the topological limit),
(b) $\mu = 1, t = \Delta = 0$ (this is the trivial phase).
Both are insulating phases; however, it is easy to see that Fig. 3.12(a) corresponds to winding of the $\mathbf{d}$-vector in the yz-plane surrounding the origin, while for Fig. 3.12(b), $\mathbf{d}$ is a constant vector in the trivial case and hence cannot wind.

Further, the energy band structure (E *vs* k) for various choices of the parameters μ, t and Δ as given in Table 3.1 are shown in Fig. 3.13 (see figure caption for details).

---

[5]Winding number denotes how many times the origin is wound around by a vector.

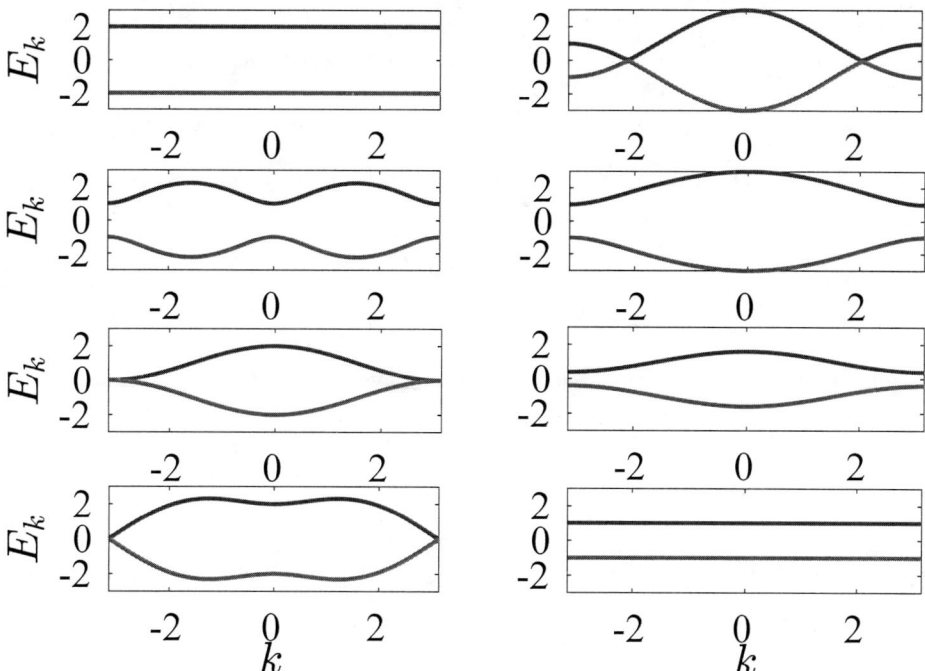

FIGURE 3.13. The eigenvalues for a two-site Kitaev model are plotted. Different choices of parameters for all the plots are to be correlated with the values of μ, t and Δ given in Table 3.1. The top-left and the bottom-right figures correspond to the topological and the trivial phases of the model. Other parameter values are for academic interest.

Both the top-left panel and the bottom-right panel of Fig. 3.13 denote flat bands and finite spectral gap everywhere in the BZ; however, the former denotes a topological gap, while the latter represents a trivial scenario. Dispersions corresponding to the other parameter values are included for academic interest.

3.2.4 Winding number

We have already discussed that the Kitaev model is invariant under particle–hole transformation, which states that an eigenstate defined by energy and momentum (E, k) will always have a partner with $(-E, -k)$. We have not made another symmetry explicit earlier, that there is also time reversal symmetry $\mathcal{T}$ defined by $\mathcal{K}$, which is simply the complex conjugation operator.[6] This means that the product of the two, that is, the particle–hole and the time reversal operations, namely $\Sigma\mathcal{T}$, known as the chiral symmetry, is a valid symmetry operation for the Kitaev model.

---

[6]In the presence of spin degree of freedom, $\mathcal{T}$ has a more complicated form, namely $\mathcal{T} = i\sigma_y\mathcal{K}$, where σ_y denotes the real spin degree of freedom.

Further, square of all of these symmetry operations, namely Σ^2 and $\mathcal{T}^2$, results in $+1$, and hence according to the tenfold classification introduced by Altland *et al.*, the symmetry belongs to the BDI class [16, 17].

Possessing a chiral symmetry enables defining a topological invariant called the winding number (we have seen this for the SSH model), which is written as

$$\nu = \frac{1}{2\pi} \int_{BZ} dk | \frac{d}{dk} h(k)|, \qquad (3.2.4.1)$$

where

$$h(k) = tan^{-1}\left(\frac{d_z}{d_y}\right) = tan^{-1}\left(\frac{\mu + 2t \cos k}{2\Delta \sin k}\right).$$

It can be easily checked that corresponding to the topological case, namely $t = \Delta = 1$ and $\mu = 0$, the winding number ν is equal to 1, while for the trivial case $\mu = 1, t = \Delta = 0$, $\nu = 0$. Same inferences have already been drawn from the behaviour of the d-vector, that is, it winds around the origin for the topological case and is merely a constant vector for the trivial case. A non-trivial topological (non-zero ν) phase is resilient against perturbations, such as disorder, defect, and so on that do not violate the chiral symmetry of the Hamiltonian.

3.2.5   Majorana fermions in the Kitaev model

Dirac equation admits both positive and negative energy solutions. While the positive energy denotes electrons, the negative solutions describe antiparticles, that is, the positrons. E. Majorana postulated that the Dirac equation can be split into two real equations where the particle and the antiparticle loose distinction. For massless and neutral particles, the corresponding antiparticles are same as the particles themselves. These are called MFs, and their existence was postulated in 1937 [31]. It is conjectured that the neutrinos denote an example of such an elementary particle whose antiparticles are thought to be the same as these particles. However, no signature of these particles has been realized in experiments so far. On the other hand, the research on MFs has gained momentum in the field of condensed matter physics and, in particular, topological superconductors, semiconductors, and so on.

Before we discuss how MFs enter into the Kitaev model, let us understand what MFs are. One can write fermion operators, namely c and $c^\dagger$, in terms of two MF operators in the following way:

$$c_i = \frac{1}{2}(\gamma_{1i} - i\gamma_{2i}) \; ; \quad c_i^\dagger = \frac{1}{2}(\gamma_{1i} + i\gamma_{2i}), \tag{3.2.5.1}$$

where γ_1 and γ_2 are two MFs defined at each site i. They are the conjugate of their own, namely

$$\gamma_1 = \gamma_1^\dagger \; ; \quad \gamma_2 = \gamma_2^\dagger. \tag{3.2.5.2}$$

The above properties can easily be verified by inverting the relations in Eq. (3.2.5.1). Thus, the MF operators represent them to be their own antiparticles. Because of this property, a single Majorana mode is never 'filled' (for example $|1\rangle$) or 'empty' (that is $|0\rangle$), unlike the way we define usual fermions. This will become clearer as we go along.

The other properties of the Majorana modes can be stated as in the following.

$$\{\gamma_1, \gamma_2\} = \{\gamma_1^\dagger, \gamma_2^\dagger\} = 0 \tag{3.2.5.3}$$
$$\gamma_1^2 = \gamma_2^2 = 1 \tag{3.2.5.4}$$

for all sites i. So the operators γ and $\gamma^\dagger$ anticommute and square to unity on their own. As one can see, even though the anticommutation relations are similar to that of fermions, they are quite distinct than the usual fermion operators and hence special. For example, the same relations for the fermion operators yield

$$\{c, c^\dagger\} = 1 \tag{3.2.5.5}$$
$$c^2 = (c^\dagger)^2 = 0. \tag{3.2.5.6}$$

Besides these, there are two other properties that we wish to highlight. The complex conjugation of the Majorana operators can be stated as

$$\mathcal{K}\gamma\mathcal{K} = \gamma \; ; \quad \mathcal{K}\gamma^\dagger\mathcal{K} = \gamma^\dagger. \tag{3.2.5.7}$$

Next, we discuss the parity of the Majorana zero modes. A few details of these modes will be discussed shortly afterwards. Consider a Kitaev chain that is sufficiently long, such that one can safely ignore any overlap between the Majoranas that exist at the two ends of the chain. The energy spectrum has two zero energy bound states. The corresponding quasiparticle operators are, say, γ_1

and γ_2. Let us combine them to form a fermion operator, using

$$c = \frac{1}{2}(\gamma_1 + i\gamma_2).$$ (3.2.5.8)

Inverting the above relation yields

$$\gamma_1 = (c + c^\dagger); \quad \gamma_2 = \frac{1}{i}(c - c^\dagger).$$ (3.2.5.9)

Thus, the Majorana operators are a superposition of the fermion creation and annihilation operators. The corresponding fermionic state can be either occupied or unoccupied. The bound states have zero energy, and hence these are degenerate. These two states differ by fermion number parity. The Kitaev model, being a mean field model for superconductors, does not conserve the particle number ($U(1)$ gauge symmetry); however, the fermion number parity remains conserved since the pairing term (that is, Δ) adds or removes particles in pairs. In a normal superconductor, the ground state always has even parity for the fermion number. With an odd number of fermions, an unpaired electron results in an energetically less favourable state.

The Kitaev chain is quite unlike a normal superconductor where the non-trivial phase hosts two ground states, each with a different parity, namely an even fermion number parity and an odd fermion number parity. We can define the fermion number parity operator as

$$P_f = 2c^\dagger c - 1 = i\gamma_1\gamma_2,$$ (3.2.5.10)

which has eigenvalues ± 1 for the two zero energy modes.

For N such Kitaev chains, there are $2N$ Majorana bound states. One can group them into N pairs, where each pair of Majorana operators γ_{2j-1} and γ_{2j} can be combined into a fermion operator c_j. The ground state now has a degeneracy of 2^N. Among these 2^N states, half of them will have one parity (say, even), and the other half will have opposite parity (odd). Now the fermion parity operator for the N-chain system is the product of the operators $2c_j^\dagger c_j - 1$ for each of the pairs, such that

$$P_f = i^N \gamma_1\gamma_2 \cdots \gamma_{2N}.$$ (3.2.5.11)

Thus, one has $\frac{1}{2}(2^N) = 2^{N-1}$ states of even and odd parity. This can be easily demonstrated in the following way. One can compute the expectation value of the fermion number parity operator $i\gamma_1\gamma_2$ within the ground state $|\phi_g\rangle$, which yields

$$\langle\phi_g|i\gamma_1\gamma_2|\phi_g\rangle = \langle\phi_g|i\gamma_1\gamma_2|\phi_g\rangle \tag{3.2.5.12}$$
$$= 1 \quad \text{for} \quad c^\dagger c = 1$$
$$= -1 \quad \text{for} \quad c^\dagger c = 0.$$

Suppose we take a simple one-site Hamiltonian, $\mathcal{H} = \mu c^\dagger c$. Writing in terms of the Majorana operators yield $\mathcal{H} = \frac{1}{2}\mu(1 + i\gamma_1\gamma_2)$. So this tells us that each fermionic site has to accommodate two MFs, as opposed to one (or zero) fermion.

Let us extend our discussion to a two-site Kitaev chain. The Hamiltonian is

$$\mathcal{H} = -\Delta(c_i^\dagger c_{i+1} + c_{i+1}^\dagger c_i + c_i c_{i+1} + c_{i+1}^\dagger c_i^\dagger), \tag{3.2.5.13}$$

where we have taken $\Delta = t$ and $\mu = 0$. There are two eigenstates of even parity and two states of odd parity corresponding to the two eigenvalues Δ and $-\Delta$ respectively. For example, the eigensolutions, $|u^e\rangle, \lambda^e$ with even parity,

$$|u_\pm^e\rangle = \frac{1}{\sqrt{2}}(1 \pm c_{i+1}^\dagger c_i^\dagger)|0\rangle \; ; \quad \text{for} \quad \lambda_\pm^e = \mp\Delta. \tag{3.2.5.14}$$

Similarly, for the odd solutions, $|u^o\rangle, \lambda^o$,

$$|u_\pm^o\rangle = \frac{1}{\sqrt{2}}(c_i^\dagger \pm c_{i+1}^\dagger|0\rangle \; ; \quad \text{for} \quad \lambda_\pm^o = \mp\Delta. \tag{3.2.5.15}$$

If we translate it into the language of MFs,

$$\mathcal{H} = -2i\Delta\gamma_{2,i}\gamma_{1,i+1}. \tag{3.2.5.16}$$

These MFs can again be combined to write down the new fermion operators, namely the d-operators[7] defined by

$$d_i = \frac{1}{\sqrt{2}}(\gamma_{1,i+1} + i\gamma_{2,i}) \; ; \quad i\gamma_{2,i}\gamma_{1,i+1} = 1 - d_i^\dagger d_i. \tag{3.2.5.17}$$

This allows the Hamiltonian to be written in terms of these d-operators.

$$\mathcal{H} = 2t d_i^\dagger d_i - t = 2t\left(d_i^\dagger d_i - \frac{1}{2}\right) = 2t\left(d_i^\dagger d_i - \frac{1}{2}\right) + 0.a^\dagger a, \tag{3.2.5.18}$$

where $a = \frac{1}{2}(\gamma_{2,1} + i\gamma_{1,N})$, which comprise the degree of freedom that is ignored in the Hamiltonian. Note that the ground state is doubly degenerate, since the states

---

[7]The d-operators here will have to be distinguished from the **d**-vectors discussed earlier.

with $\langle a^\dagger a \rangle = 0$ and 1 have the same energy. Suppose N is odd, the state $|0\rangle$ has an even parity, which means that it has even number of a fermions. In the same way, the other ground state, namely $|1\rangle$, has one fermion and is an odd-parity state. Thus, these two states are quite intriguing in the sense that they can be considered as the superposition of two MFs that reside at the two edges of the chain. The occurrence of these Majorana zero modes and their degenerate nature are artefacts of the topological properties of the model. Evidently, the Hamiltonian commutes with $\gamma_{1,i}$ and $\gamma_{2,i+1}$, that is,

$$\gamma_{1,i}|u^e_+\rangle = \frac{i}{2}(c^\dagger_i + c^\dagger_{i+1})|0\rangle; \;\; \gamma_{2,i+1}|u^e_+\rangle = \frac{1}{2}(c_{i+1} + c_i)|0\rangle = -i\gamma_{1,i}|u^e_+\rangle, \;\; (3.2.5.19)$$

which says that the states are identical and differ by only a phase factor.

Let us now consider a chain of N sites (in Fig. 3.14, we show $N = 4$). At each site, there is a fermion, and hence there are two Majorana modes, namely γ_{2j-1} and γ_{2j}. These are called the domino tiles for the Majorana. There are 10 Majorana modes denoted by $\gamma_1 \ldots \gamma_{10}$. The Kitaev Hamiltonian written in terms of the MFs,[8]

$$\mathcal{H} = -\frac{i\mu}{2}\sum_{j=1}^{N}(1 + i\gamma_{1,j}\gamma_{2,j}) - \frac{i}{4}\sum_{j=1}^{N-1}\left[(\Delta + t)\gamma_{2,j}\gamma_{1,j+1} + (\Delta - t)\gamma_{1,j}\gamma_{2,j+1}\right].$$

$$(3.2.5.20)$$

This has a formal similarity with the two coupled SSH chains with an onsite energy, $-\frac{i\mu}{2}$, and hopping terms as $i(\Delta + t)$ and $i(\Delta - t)$. However, we shall not explore this symmetry further.

One way of pairing the Majoranas is to pair them in the same block, and the other would be to pair them across each block. In the first case, the Hamiltonian

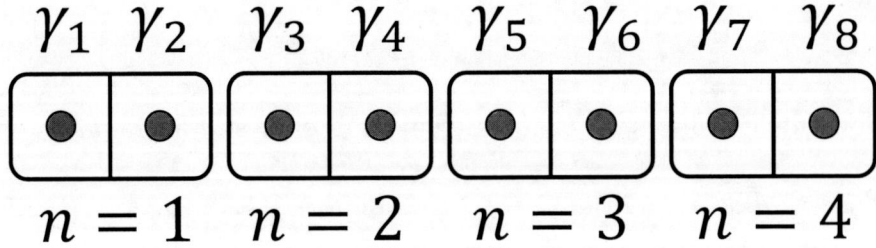

FIGURE 3.14. The domino model is sketched for $N = 4$.

---

[8]We discontinue the usage of the symbol 'i' for denoting the site index because of the imaginary i being present in the Hamiltonian, and use j instead.

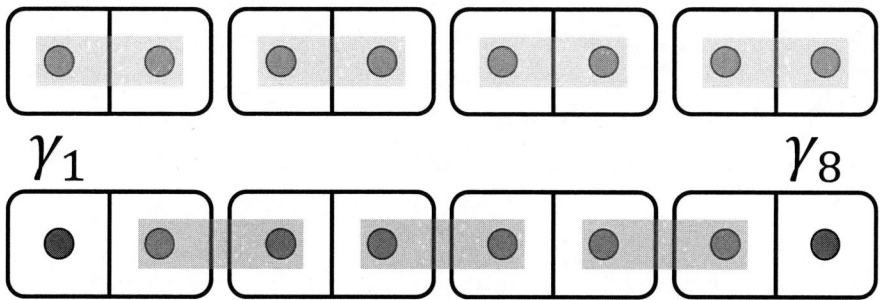

FIGURE 3.15. The trivial and the topological domino models are demonstrated.

has only onsite energy, that is, μ for the fermions. This yields a Hamiltonian

$$\mathcal{H} = -\frac{i\mu}{2} \sum_{j=1}^{N} \gamma_{1,j} \gamma_{2,j}. \tag{3.2.5.21}$$

This situation clearly subscribes to the trivial case, where $\Delta = t = 0$ and only $\mu \neq 0$. Thus, there are no unpaired Majorana, and the spectrum is gapless with excitation energies $\pm\mu$. Correspondingly, it has no edge modes, and the system is a trivial insulator. However, the other scenario that pairs two Majoranas across the domino tiles yields a Hamiltonian,

$$\mathcal{H} = it \sum_{j=1}^{N-1} \gamma_{2,j} \gamma_{1,j+1}. \tag{3.2.5.22}$$

Clearly, this leaves the two Majorana at the edges to be excluded in $\mathcal{H}$. Also we may recall that this Hamiltonian corresponds to $t = \Delta = 1$ and $\mu = 0$ which denotes the topological limit of the Kitaev model.

The first case discussed here corresponds to the trivial topology (see upper panel of Fig. 3.15), where the intra-cell pairing of the Majoranas yields no zero modes. However, the latter (shown in the lower panel of Fig. 3.15) hosts Majorana zero modes and thus denotes a topological phase of the system. It is rewarding to realize that these two cases for the pairing of Majoranas in a domino model correspond to zero modes in the Kitaev model that we have discussed earlier.

An intuitive way to arrive at the zero energy Majorana modes in the Kitaev chain can be seen by exploring the Kitaev Hamiltonian. For that purpose, we shall write down the Kitaev Hamiltonian once again in real space.[9] We begin with a

---

[9]This real space representation is distinct from the one written in Eq. (3.2.2.1) and is obtained by discretizing the momentum operator on a chain.

small k version of the Hamiltonian, where the $-2t \cos k$ in diagonal term (see Eq. 3.2.2.10) is replaced by $2t$ (this also necessitates the off-diagonal terms to be written as $2i\Delta k$). Hence, we take a small deviation of the chemical potential μ from its value where the phase transition occurs from a topological to a trivial phase, namely $\mu = \pm 2t$. As a specific case, assume

$$\mu = -2t + m, \tag{3.2.5.23}$$

where m is a small and positive quantity. This yields the **d**-vector to have a form

$$\mathbf{d}(k) = (0, 2\Delta k \sigma_y, m\sigma_z). \tag{3.2.5.24}$$

We write k as $\frac{\partial}{\partial x}$ (the chain is placed along the x-direction) and assume m to be a smoothly varying quantity in x, that is, $m(x)$, which changes sign across the transition. This allows us to write the Hamiltonian as

$$\mathcal{H} = 2i\Delta\sigma_y \frac{\partial}{\partial x} + m(x)\sigma_z. \tag{3.2.5.25}$$

The time-independent Schrödinger equation corresponding to $E = 0$ for the Hamiltonian can be written as

$$\left(2i\Delta\sigma_y \frac{\partial}{\partial x} + m(x)\sigma_z\right) \phi(x) = 0, \tag{3.2.5.26}$$

where $\phi(x)$ is the eigenfunction. Solving the above equation, one gets two solutions (corresponding to the 2×2 structure of the Pauli matrices) as

$$\phi_\pm(x) = exp\left(\pm \int \frac{m(x)}{2\Delta}dx\right) \phi_\pm(0), \tag{3.2.5.27}$$

where $\phi_\pm(0)$ refers to the eigenfunction of σ_x, that is,

$$\phi_\pm(0) = \begin{pmatrix} 1 \\ \pm 1 \end{pmatrix}.$$

The $\pm$-sign yields an unphysical situation as it diverges at $x = \infty$. However, there is a solution corresponding to $m(x) = 0$. This says that the zero energy mode decays exponentially in either direction with respect to $x = 0$. We can invoke any functional behaviour of $m(x)$; however, the behaviour of the zero mode remains unaltered. Suppose, we assume $m(x)$ to change sign at $x = 0$ and flat otherwise (see Fig. 3.16), the bound state, $\phi_-(x)$, assumes the form

$$\phi_-(x) = exp\left(-|m|x\sigma_x\right) \phi_-(0). \tag{3.2.5.28}$$

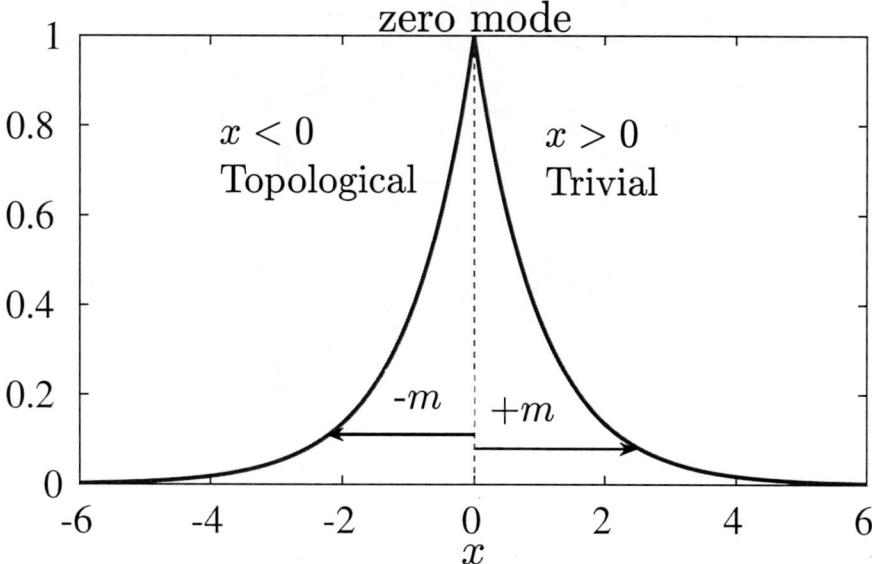

FIGURE 3.16. The spatial dependence of the wavefunction is shown. It falls off exponentially on either side of $x = 0$, the left of which denotes a topological phase, while the right represents a trivial phase.

3.2.6   Energy spectrum of N-site Kitaev model

In order to obtain the energy spectrum of the Kitaev model, we can use Eq. (3.2.2.11), which was originally derived for a two-site Kitaev chain.[10] However, for an N-site system, the same expression can still be used with N modes, that is, N k-values equally spaced between $-\pi$ and $+\pi$.

It is also possible to solve Eq. (3.2.5.21) and Eq. (3.2.5.22) in the basis of MFs in real space. For a chain consisting of N-sites, the Hamiltonian in Eq. (3.2.5.21) (or Eq. 3.2.5.22) yields N eigenvalues, which can be mapped on to N distinct k-values that belong to the 1D BZ. The spectrum corresponding to the topological case (Eq. 3.2.5.22) is plotted in Fig. 3.17 as a function of the chemical potential, μ. There are two degenerate zero modes that persist till $\mu = 2t$, beyond which these modes merge into the bulk. Hence, for $\mu > 2t$, one obtains a trivial phase. For the trivial case, the spectrum is gapped for all values of μ (see Fig. 3.18). The topological protection of the zero modes is discussed below.

---

[10]We have written it for two sites in Eq. (3.2.2.1).

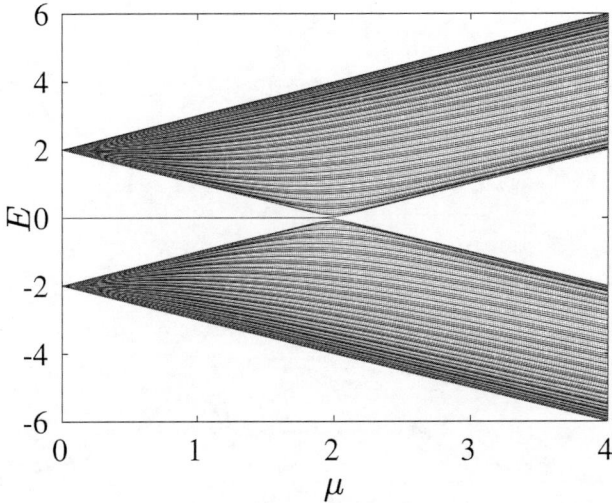

FIGURE 3.17. The spectrum for a Kitaev model is plotted for $t = \Delta = 1$. The zero modes persist till $\mu = 2t$, beyond which a trivial band gap opens up.

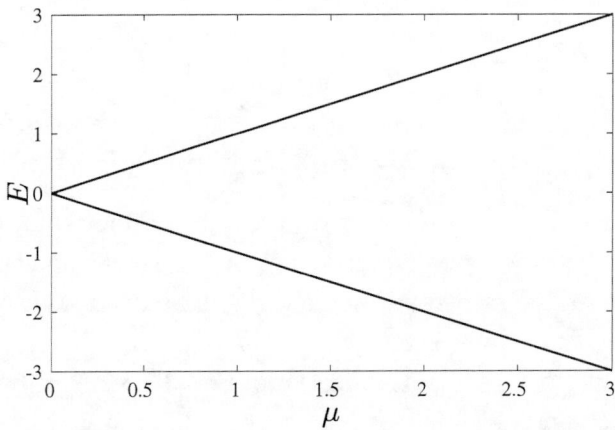

FIGURE 3.18. The spectrum for a Kitaev model is plotted for $t = \Delta = 0$. A band gap be seen for all values of μ.

3.2.7   Topological properties of the Majorana modes

It may seem that the topological phase of the Kitaev chain is an artefact of setting the chemical potential μ equal to zero thereby disconnecting the two Majorana modes at the two edges. So a natural question is: if we change μ slightly from a value zero, do the Majorana modes disappear by coupling to the rest of the chain? If this is true, then it results from a very carefully controlled model. In reality, this is not true. If we increase μ, it can be easily checked that till $\mu \leq 2t$, the zero modes (two of them) stay together, and they only split for $\mu > 2t$. One can see the same behaviour for negative values of the chemical potential, that is, the zero modes stick together at $E = 0$ for $\mu \geq -2t$. At $\mu = \pm 2t$, the system becomes critical and the bulk gap closes, whence the system ceases to be topological. Thus, the Majorana modes are protected as long as there is a gap in the bulk spectrum.

The Majorana modes are indeed protected by the particle–hole symmetry of the Hamiltonian. For $\mu = 0$ (and $\Delta = t \neq 0$), there are equal number of particle-like (corresponding to $E < 0$) and hole-like states (for $E > 0$). Further, there are two states at zero energy that are inseparable, since if that were untrue, for example, if they are separated, then that will cause imbalance in the number of states below and above $E = 0$ (Fermi energy). The only possibility is to couple the two unpaired Majorana modes to each other; however, such a coupling is impossible because of the large spatial distance between them. The only way to facilitate a coupling is to close the bulk gap, which precisely happens for $\mu = \pm 2t$ where the zero modes disappear into the bulk.

3.2.8   Experimental realization of the Kitaev chain

From the discussion so far, it is clear that there are intriguing possibilities of topological features emerging out depending on parameters of the Hamiltonian. The zero energy mode in the topological regime is a coherent superposition of the two Majoranas, and has opposite fermion number parity. However, the system may look rather unphysical. Most crucial hindrance is provided by freezing out the spin degrees of freedom. How do we do that? For argument's sake, we can include a strong magnetic field that will polarize all the spins of the electrons in the same direction of the field. However, that brings us to the conflict of superconductivity coexisting with a strong magnetic field. Further, the availability of p-wave superconductors in nature is infrequent. Additionally, not to forget that a mean field theory of superconductivity is susceptible to large fluctuations in one dimension which may prevent stabilization of the superconducting state.

Nevertheless, the seminal works of Fu and Kane [32, 33] have made it abundantly clear that a Kitaev chain can be experimentally realized in a variety of systems. The key ingredients are proximity induced superconductivity and spin–orbit coupling (SOC) where the latter boosts experimental search for exploring SOC in superconductors. From undergraduate level quantum mechanics, it is known that the orbital angular momentum ceases to be a good quantum number in the presence of SOC, and thus it raises the possibility of p-wave superconductivity (as an admixture to the s-wave pairing) which entails a spin polarized pairing, and can possibly be induced by proximity effects. For more detailed discussions, one may refer to the papers in Refs. [34, 35].

As discussed earlier that the Majorana bound states are superposition of particle and hole excitations at $E = 0$ which is trapped in the bulk gap. Thus, two Majorana bound states encode a non-local qubit that is robust against local perturbations, such as decoherence and thus constitute essential elements for topological quantum computation. The Majorana bound states can be experimentally probed by Andreev reflection. Suppose in an experiment with a semiconducting wire placed on a p-wave superconductor (or even an s-wave superconductor is deposited on the surface of a topological insulator whose surface states are denoted by the likes of Dirac fermions), a metallic lead couples to the Majorana modes by tunnelling of electrons. The MFs can induce resonant Andreev reflection from the lead to the superconductor [36, 37]. Usual Andreev reflection converts an electron into a hole in the same lead; however, a crossed Andreev reflection non-locally converts an electron excitation into a hole excitation into a different lead. Equivalently, it splits a Cooper pair and distributes them into two leads, which happens at very low excitation energies.

3.3   Detecting Majoranas: 4π periodic Josephson junctions

Consider two Kitaev chains in their topological phases such that the respective ends contain MFs. The arrangement can be joined to form a Josephson junction (JJ). One can write a tunnelling Hamiltonian of the form,

$$\mathcal{H}_{JJ} = t c_L^\dagger c_R + t^* c_R^\dagger c_L, \tag{3.3.1}$$

where t and t^* denote the hopping amplitudes between the left and right ends and vice versa respectively. Projecting the electron operators onto the Majoranas,

$$c_L \simeq u_L \gamma_L \; ; \; c_R \simeq i u_R \gamma_R, \tag{3.3.2}$$

where γ_L and γ_R are Majorana operators, and u_L and u_R are the (real) wavefunctions associated with the Majoranas in the left and right chains respectively. Thus, the Hamiltonian becomes

$$\mathcal{H}_{JJ} = i(t + t^*) u_L u_R \gamma_L \gamma_R. \tag{3.3.3}$$

One can choose a gauge such that a phase difference of ϕ exists between the two chains. The hopping amplitude picks up this phase via

$$t = t_0 e^{i\phi/2} \quad (t_0 : \text{real}). \tag{3.3.4}$$

This means that each of the hopping terms picks up a phase $\phi/2$. It yields the form of the Hamiltonian as

$$\mathcal{H}_{JJ} == it_0 \left(e^{i\phi/2} + e^{-i\phi/2} \right) u_L u_R \gamma_L \gamma_R = 2t_0 \cos(\phi/2) u_L u_R P, \tag{3.3.5}$$

where

$$P = i\gamma_L \gamma_R$$

denotes the parity operator. Thus, the energy spectrum is given by

$$E_{\pm} = \pm 2t_0 \cos(\phi/2) u_L u_R, \tag{3.3.6}$$

where the $\pm$-sign denotes the eigenvalues of the parity operator.

Owing to the presence of the phase $\phi/2$ in Eq. (3.3.6), the energy E is periodic with a period 4π (see Fig. 3.19). A phase difference of 2π switches from one eigenvalue to another, that is, from E_+ to E_- or vice versa. For a given parity eigenvalue (say, $+1$), this results in a 4π periodic Josephson current, and is in sharp contrast with the usual 2π periodic modulation of the current. Thus, the consequence of the presence of Majoranas can be tested by measuring the Josephson current obtained from the energy spectrum via

$$I(\phi) = 2e\frac{dE}{d\phi}. \tag{3.3.7}$$

The magnitude of the current remains the same, albeit it comes with opposite signs for different fermion parities.

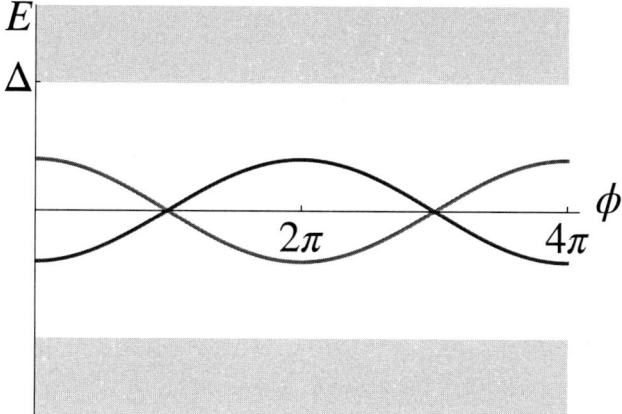

FIGURE 3.19. The energy spectrum E corresponding to Eq. (3.3.6) is shown. E is clearly 4π periodic for a fixed parity.

Further, the energies corresponding to the two parity states cross zero for $\phi = \pm\pi$ and are degenerate. In fact, it can be interpreted that these subgap states (energy being lower than the superconducting gap Δ for the system to be in the topological phase) undergo a quantum phase transition at $\phi = \pm\pi$ at which the fermion parity of the many body ground state switches. Summarizing, the 4π periodicity produces an ac current at half the conventional Josephson frequency, and thus the scenario advocates unconventional or topological superconductivity.

There are alternate methods to experimentally realize the Majorana modes which employ the concept of Andreev reflection. However, a detailed analysis of it is beyond the scope of this book. We shall include a very brief note on it in the following.

Consider a superconducting Kitaev chain in the topological phase that is connected to a metallic lead. Tunnelling of the charge carriers into the superconducting region is suppressed owing to a gap (of magnitude 2Δ) at the Fermi level. However, Andreev reflection is possible across the junction. In a few words, when an electron arrives at the interface between a metal and a superconductor, three processes simultaneously occur. They are:

(i) a normal specular (mirror-like) reflection of the incident electron.
(ii) A retro-reflection of a hole (tracing the path of the incident electron).
(iii) Injection of a Cooper pair into the superconducting side of the junction.

In the presence of a bias voltage, eV, the Andreev current at an energy ω depends on the tunnelling amplitude, the Fermi distribution functions of the incoming electrons, namely $f(\omega - eV)$, that of the reflected holes, $[1 - f(\omega + eV)]$, and the scattering amplitude A for the corresponding process. Thus, the Andreev current which is an experimentally measurable quantity can be obtained as

$$I \propto 2e \int d\omega |A|^2 f(\omega - eV)[1 - f(\omega + eV)]. \tag{3.3.8}$$

The resonance feature of the integrand at $\omega = 0$ signals the presence of Majorana bound states at the junction.

Having discussed topology in two 1D tight-binding models, namely a dimerized chain (SSH model) and a system comprising of spinless fermions with p-wave superconducting correlations (Kitaev model), in the following section, we introduce a quasi-1D system, such as a ladder and explore its topological properties.

3.4 Creutz ladder

Having discussed two paradigmatic models in 1D, we focus towards the topological aspects of a quasi-1D system, namely a ladder in the subsequent discussion. A Creutz ladder [38] is a quasi-1D system that hosts robust topological edge states. The model is characterized by a horizontal, a vertical and a diagonal hopping term. These form the parameter space of the system, and in a certain region of the parameter space, it demonstrates topological properties via the presence of robust edge states. We analyze the symmetries and the phase diagram of the Creutz ladder. There is also a magnetic field that threads the ladder and yields an extra degree of freedom in terms of rendering a complex phase to the hopping amplitudes. Interestingly, in spite of the presence of a magnetic field, the time reversal symmetry is preserved for the system. Quantum interference yields robust edge modes which do not get easily destroyed and are expected to find their use in quantum information.

The Creutz ladder consists of two rungs of lattice sites, and is characterized by diagonal, horizontal and vertical hopping amplitudes. Additionally, a magnetic field penetrates the ladder in a plane perpendicular to it. We use a Landau gauge to characterize the magnetic field. As a result, the horizontal hopping amplitudes carry a phase along with them. This complex phase leads to destructive interference of the hopping amplitudes and hence show localization of the particles for certain regions of the parameter space. This model being quasi-1D, it is difficult to place it into the conventional tenfold classification of symmetries.

It is interesting to note that zero modes in the Creutz ladder result from the dual effects of Aharonov–Bohm caging and topology. In the presence of this dual protection, edge modes are robust even for small systems. There is a debate over whether the Creutz Ladder belongs to the symmetry class AIII or BDI [40] (see the Appendix section in this chapter for a description of these symmetry classes and the corresponding topological invariants). In the presence of an external magnetic field, the presence of a time reversal symmetry seems mysterious. However, it can be explained by correlating the two rungs of the Creutz ladder to the two spin components of a spin-$\frac{1}{2}$ fermion. In that case, time reversal operator which is only responsible for interchanging the two rungs of the ladder can be associated with the conventional time reversal.

The band structure of the Creutz ladder shows a flat band for the case of no vertical hopping, that is, $R = 0$ with equal magnitudes of horizontal (L) and diagonal (D) hopping terms. This implies that the group velocities of the resulting states are all zero. In the open boundary condition, this leads to

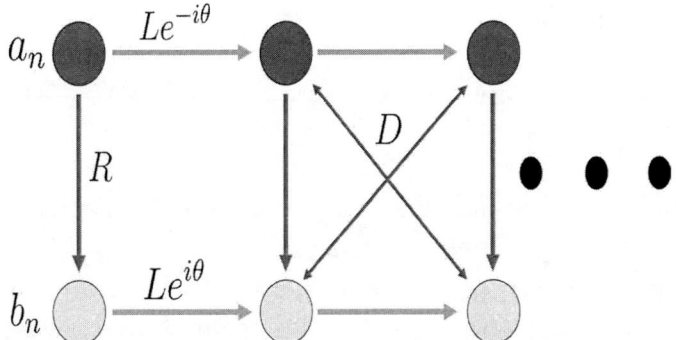

FIGURE 3.20. Pictorial representation of the original Creutz ladder.

states that are completely localized at the edges. Such phenomena have been experimentally realized in certain ultracold atoms and photonic lattices [41]. The complete localization of the edge modes at the boundaries, however, gets disturbed if we move slightly in the parameter space, and the edge modes can become exponentially localized with a little of it penetrating into the bulk. We study the model in its parameter space defined by L, R, D and θ (see Fig. 3.20) and plot the corresponding spectral features. The behaviour of the edge modes is investigated, along with obtaining a phase diagram in the $R - \theta$ space which shows a topological phase transition.

It will be helpful to a priori state the main results. The model shows topological properties with robust edge modes for $R < 2$ in units of $L = D = 1$ and $\theta = \frac{\pi}{2}$. At $R = 2$ (keeping all the other parameters unchanged), it undergoes through a gap-closing transition, and subsequently becomes a trivial insulator for $R > 2$.

3.4.1   The Hamiltonian

The Creutz ladder [38] deals with the Hamiltonian,

$$\mathcal{H} = -\sum_n L(e^{i\theta} a_n^\dagger a_{n+1} + e^{-i\theta} a_{n+1}^\dagger a_n + e^{-i\theta} b_n^\dagger b_{n+1} +$$

$$e^{i\theta} b_{n+1}^\dagger b_n) + \quad\quad (3.4.1.1)$$

$$D(a_n^\dagger b_{n+1} + b_{n+1}^\dagger a_n + a_{n+1}^\dagger b_n + b_n^\dagger a_{n+1}) +$$

$$R(a_n^\dagger b_n + b_n^\dagger a_n).$$

Here L, D and R denote the horizontal, diagonal and vertical hopping amplitudes respectively. θ is the phase introduced by the external magnetic field. If ϕ denotes the total flux through each plaquette, then $2\theta = \frac{\phi}{\phi_0}$, where ϕ_0 denotes the magnetic flux quanta. The energy spectrum can be computed in the real space for a finite ladder, and the corresponding results along with localization of the edge modes are discussed later.

We Fourier transform this Hamiltonian using

$$a_n = \sum_k a_k e^{-ikx_n}, \quad a_n^\dagger = \sum_k a_k^\dagger e^{ikx_n}. \tag{3.4.1.2}$$

Here a_n and $a_n^\dagger$ represent the annihilation and creation operators in real space. n corresponds to the nth rung of the ladder. The Hamiltonian in the momentum space now reads as

$$H(k) = 2Lcos(k)cos(\theta)\sigma_0 + 2Lsin(k)sin(\theta)\sigma_z +$$
$$(R + 2Dcos(k))\sigma_x. \tag{3.4.1.3}$$

The basis used in this case is (a_k, b_k).

3.4.2   Symmetries of the Hamiltonian

It is important at this point to understand the symmetries of the system. The model has an inherent inversion symmetry given by the relation

$$\sigma_x \mathcal{H}(k)\sigma_x = \mathcal{H}(-k).$$

Furthermore, it possesses a chiral symmetry given by

$$\sigma_y \mathcal{H}(k)\sigma_y = -\mathcal{H}(k)$$

corresponding to $\theta = \frac{\pi}{2}$. The chiral symmetry is broken for other values of the phase, θ. As previously mentioned that there is a mysterious time reversal symmetry in the model which is maintained throughout the parameter space in spite of the presence of a magnetic field. It is given by

$$\sigma_x \mathcal{H}^*(k)\sigma_x = \mathcal{H}(-k).$$

Lastly, a particle–hole symmetry exists in the system for $\theta = \frac{\pi}{2}$, which may be expressed as

$$\sigma_z H^*(k)\sigma_z = -H(-k).$$

Accordingly, the Hamiltonian at the flat band point can be placed in the BDI or AIII symmetry class depending on whether the TRS and PHS are respectively both present and absent.

3.4.3   Energy spectrum of the Creutz ladder

The corresponding energy bands in the k-space show several interesting features (see Fig. 3.21). The first being the flat band corresponding to $R = 0$ with $L = D = 1$ and $\theta = \frac{\pi}{2}$ (see Fig. 3.21a). As R is increased to 1 (leaving other parameters unchanged), the bands become dispersive with spectral gaps observed at $k = \pm\pi$ (see Fig. 3.21b). At $R = 2$ (see Fig. 3.21c), the gap closes and later reopens again for $R = 2.2$ (Fig. 3.21d). Thus, the systems undergo through a gap-closing transition. A priori, the gap is topological for $R < 2$ and trivial for $R > 2$. We show this in the next section.

It is instructive (and of academic interest) to examine the cases for other values of θ, where the system looses chiral symmetry (see Fig. 3.22). For example, consider $\theta = 0$ (Fig. 3.22a) and $\theta = \pi$ (Fig. 3.22b). It is observed that one band is dispersive and the other one is completely flat. The flat band for $R = 2, L = 1, D = 1$ and $\theta = \pi$ lies at $2L$ ($L = 1$ here), while for $\theta = 0$ it is at $-2L$.

It is also instructive to examine the scenario in real space. The numeric computation of the Hamiltonian in Eq. (3.4.1.1) has been done on a finite lattice consisting of 50 unit cells (that is, 100 sites) along the horizontal (leg) direction, and the results are shown in Fig. 3.23. In Fig. 3.23(a), there are zero modes for $R = 1$, which are absent in Fig. 3.23(b) obtained for $R = 2$.

Further, the localization of the zero energy modes can be checked by plotting the square of the wavefunction at each of the sites of a Creutz ladder, which we show in Fig. 3.24. Corresponding to $R = 0$ and $R = 1$ (as long as $|R|$ remains at values lower than 2), the zero modes are localized at the boundaries of the ladder, while they are extended for $R \sim 2$.

3.4.4   Topological phases and winding number

Both the symmetry classes BDI and AIII host topological phases in 1D, which are characterized by a $\mathbb{Z}$-type topological invariant. Writing the Hamiltonian in the massless Dirac form as

$$\mathcal{H}(\mathbf{k}) = d_0\sigma_0 + \mathbf{d}(\mathbf{k}).\sigma, \tag{3.4.4.1}$$

where the expressions for d_0, d_x and d_z are given by

$$d_0 = 2L\cos k\cos\theta,$$

$$d_x = R + 2D\cos k$$

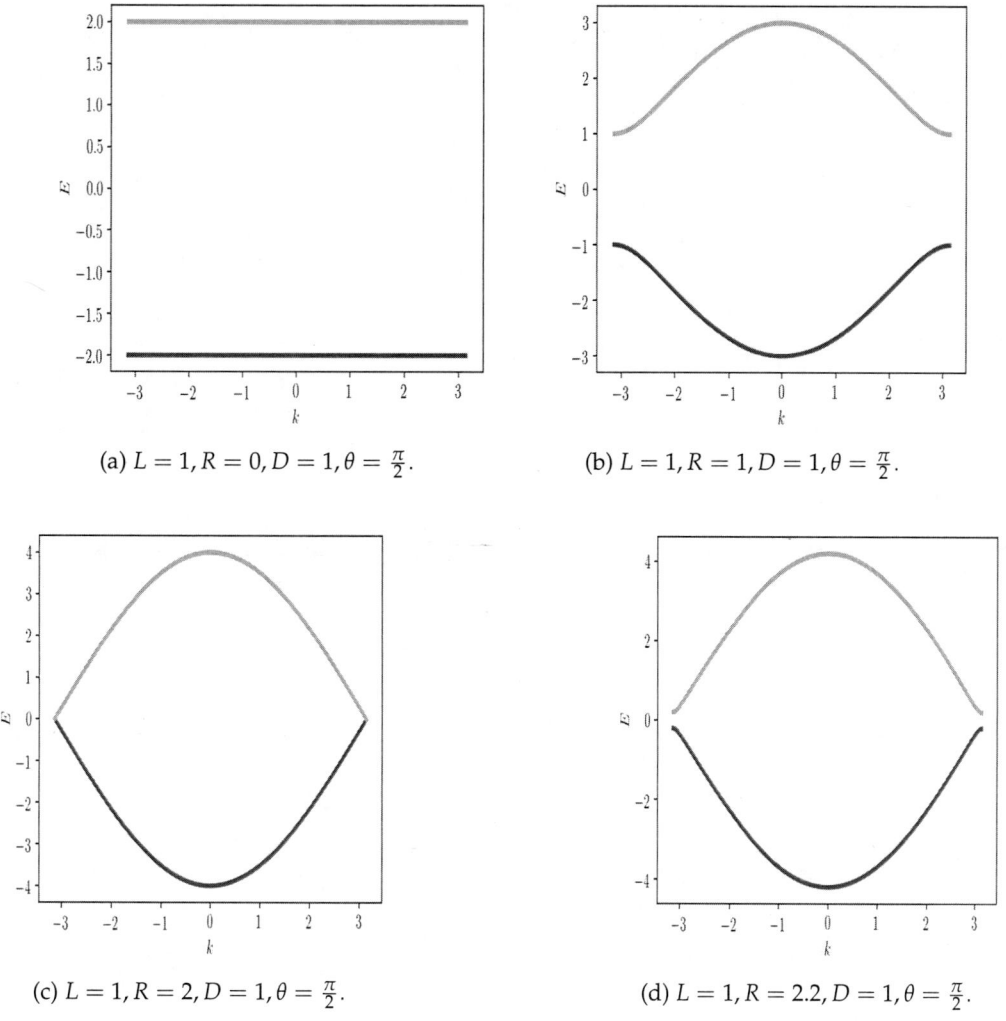

(a) $L = 1, R = 0, D = 1, \theta = \frac{\pi}{2}$.

(b) $L = 1, R = 1, D = 1, \theta = \frac{\pi}{2}$.

(c) $L = 1, R = 2, D = 1, \theta = \frac{\pi}{2}$.

(d) $L = 1, R = 2.2, D = 1, \theta = \frac{\pi}{2}$.

FIGURE 3.21. The band structures are shown for the Creutz ladder. (a) shows the flat band dispersion where all group velocities are zero. Robust edge states are found at this point under open boundary conditions. (b) represents the point $L = 1, R = 1, D = 1, \theta = \frac{\pi}{2}$. Plots in (c) show the dispersion at $R = 2$ (the band closing point) and (d) denotes the same for $R = 2.2$, where the band gap reopens, which, however, yields an insulating gap.

and

$$d_z = 2L \sin k \sin \theta,$$

where σ denotes the Pauli matrices and σ_0 is 2×2 identity matrix. We associate the topological invariant with the winding in the d_x–d_z plane. It is to be understood

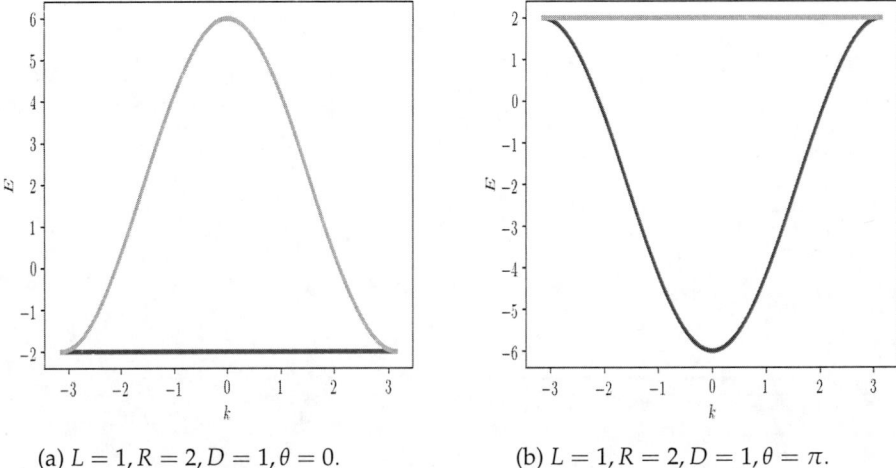

(a) $L = 1, R = 2, D = 1, \theta = 0$.   (b) $L = 1, R = 2, D = 1, \theta = \pi$.

FIGURE 3.22. (a) shows dispersion for $\theta = 0$. (b) shows the same for $\theta = \pi$. Other parameters are indicated in the figures. One band is dispersive, while the other one is flat. For other values of R, there will a gap at $k = \pm\pi$.

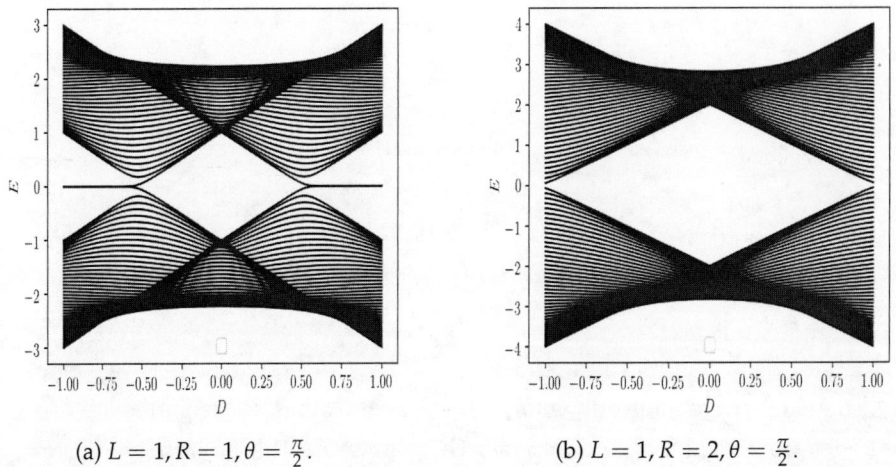

(a) $L = 1, R = 1, \theta = \frac{\pi}{2}$.   (b) $L = 1, R = 2, \theta = \frac{\pi}{2}$.

FIGURE 3.23. (a) shows the dispersion in real space for $R = 1$ as a function of the diagonal hopping, D. The edge modes are visible. (b) shows the same for $R = 2$, where there are no edge modes.

that the winding number is a highly robust quantity, which cannot be modified by smooth deformation of the system. The prescription for calculating the winding number is

$$\nu = \frac{1}{2\pi} \int_0^{2\pi} \frac{d_z d(d_x) - d_x d(d_z)}{d_x^2 + d_z^2} dk. \qquad (3.4.4.2)$$

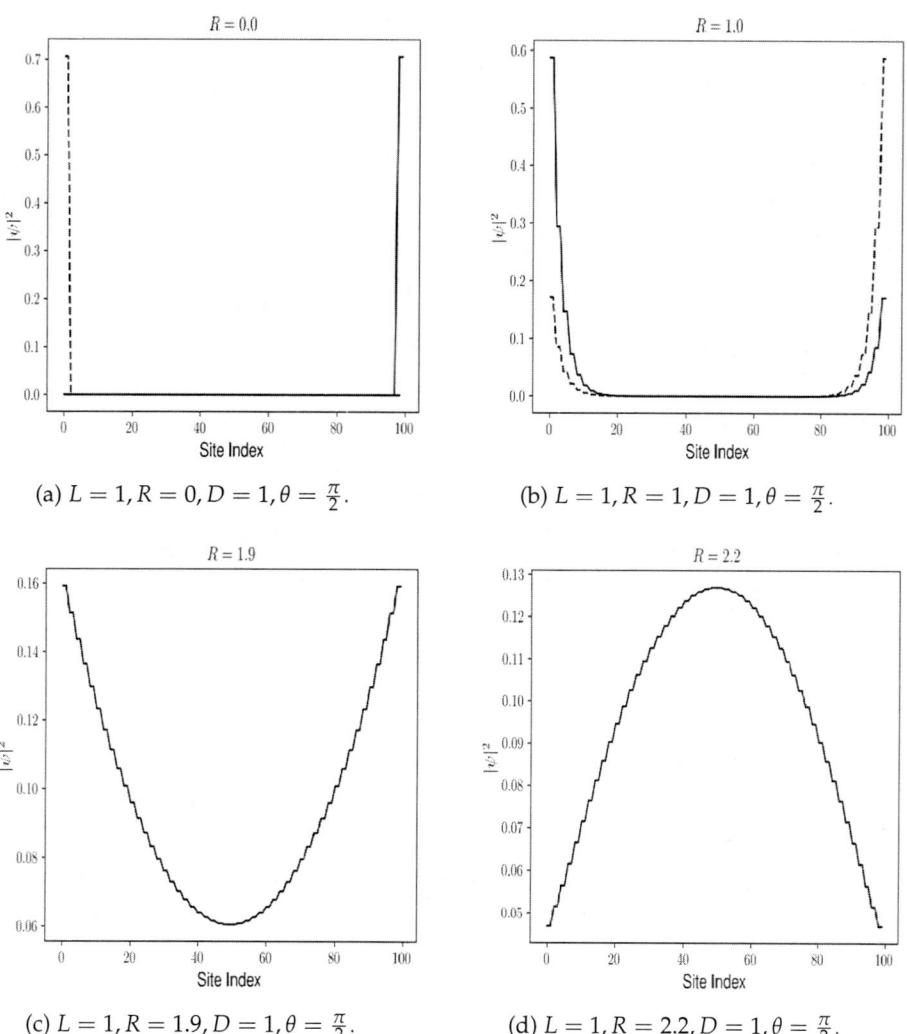

(a) $L = 1, R = 0, D = 1, \theta = \frac{\pi}{2}$. (b) $L = 1, R = 1, D = 1, \theta = \frac{\pi}{2}$.

(c) $L = 1, R = 1.9, D = 1, \theta = \frac{\pi}{2}$. (d) $L = 1, R = 2.2, D = 1, \theta = \frac{\pi}{2}$.

FIGURE 3.24. The probability amplitude $(|\psi|^2)$ are plotted for $R = 0$ in (a), $R = 1$ in (b), $R = 1.9$ in (c) and $R = 2.2$ in (d). The localization of the edge modes is clear from (a) and (b), while they are extended for (c) and (d).

Here $\nu = \pm 1$ for the topological phase and 0 for the trivial phase. The real space band structure shows that the system is trivial for $|R| > 2$ (in unit of L) and topological for $|R| < 2$. However, the edge modes show complete localization at the edges only at the flat band point given by $R = 0$ and $\theta = \frac{\pi}{2}$. Away from this point, there is no chiral symmetry or particle–hole symmetry. Only the inversion symmetry protects the edge modes, though they show exponential decay within the bulk in this situation. The winding of the origin in the d_z–d_x plane is shown

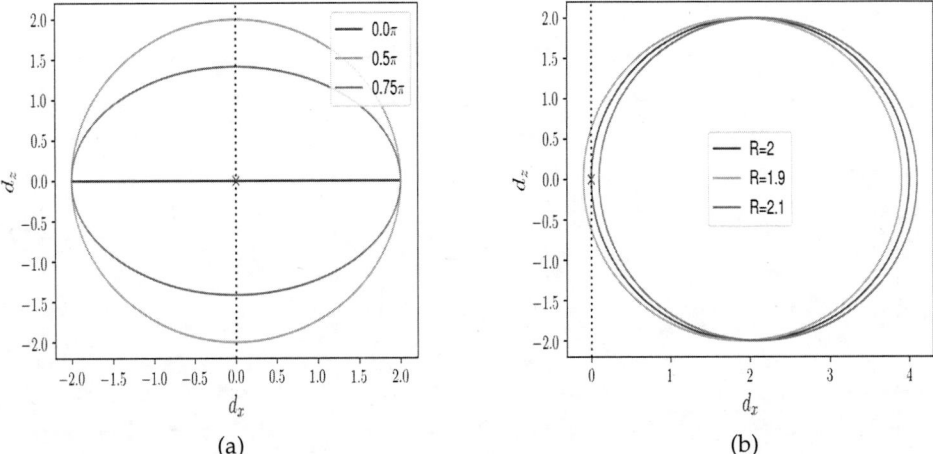

(a) (b)

FIGURE 3.25. (a) shows the winding of origin in the d_z–d_x plane for different values of θ and $R = 0, D = 1, L = 1$. The origin is wound in every case implying that all phases are topologically non-trivial. (b) shows winding for different values of R corresponding to $\theta = \frac{\pi}{2}, D = 1, L = 1$. The origin is wound only when $R < 2$. Above this, the system becomes topologically trivial as is shown by $R > 2$, that is, $R = 2.1$ plot.

in Fig. 3.25. As can be clearly seen that for $R = 0$ (no vertical hopping, that is, the flat band point), the components of the vector d winds the origin for all values of θ (Fig. 3.25a). Whereas, for $R > 2$, $|d| = \sqrt{d_z^2 + d_x^2}$ does not enclose the origin (Fig. 3.25b), thereby indicating formation of a topologically trivial phase.

Finally, we show the phase diagram for the winding number ν in the $R - \theta$ plane in Fig. 3A.1, which clearly shows three distinct phases. $\nu = \pm 1$ corresponds to the topologically non-trivial phases, while $\nu = 0$ denotes a trivial phase. It is interesting to note that for a given value of R, such that $|R| < 2$, the phase characterized by $\nu = +1$ transforms to $\nu = -1$ through an arguably trivial region as θ is varied (not shown here). It is also to be noted that for θ values other than $\theta = \frac{\pi}{2}$, the winding number is '*barely*' preserved due to the presence of the inversion symmetry, since both the chiral symmetry and the particle–hole symmetry cease to exist.

3A Appendix

3A.1 Periodic table of topological materials: Tenfold classification

The tenfold symmetry classification of topological matter involves categorizing topological insulators and superconductors on the basis of the presence or

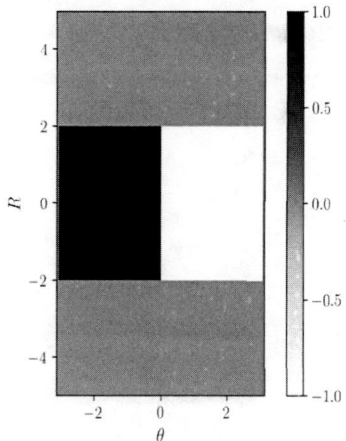

FIGURE 3A.1. The phase diagram of the Creutz ladder is shown in the $R-\theta$ plane. There are three phases $\nu = \pm 1$ (shown by black and white) and $\nu = 0$ (shown by grey). Non-zero values of ν correspond to topological phases, while $\nu = 0$ denotes a trivial phase.

absence of three discrete symmetries. These symmetries include the time reversal symmetry, particle–hole (or equivalently charge-conjugation) symmetry and chiral symmetry. In order to decide which class a material falls into, we investigate which of the aforementioned symmetries the material hosts.

It is important to understand the inherent physical meaning of these non-spatial symmetries in mathematical terms. We have done this earlier in Section 3.2, but wish to repeat here some of the essential properties for the discussion to be self-sufficient in this appendix. The time reversal operator has the effect of reversing the motion of a particle, that is, it transforms $\mathbf{k}$ to $-\mathbf{k}$. If the evolution of a system under the action of time reversal operator remains the same, the system is called time reversal symmetric. Mathematically, the time reversal operator is represented by K (for spinless) or $i\sigma_y K$ (for spinor systems). Here, K represents complex conjugation operator and σ_y is the y-component of the Pauli matrix. The charge-conjugation operator, as the name suggests, physically represents replacing a particle by its conjugate within the system. If under such a replacement, the system remains invariant, we call it to be particle–hole symmetric. Importantly, the time reversal symmetry and the particle–hole symmetry are anti-unitary symmetries which square to ± 1. Lastly, there are systems that are neither symmetric under time reversal nor particle–hole symmetries. However, if the motion within the system is reversed and the particles are replaced by their conjugates simultaneously, invariance is re-established. Such systems are purely chiral systems. Chiral symmetry is represented as the product of the time reversal and the particle–hole symmetry. It is difficult to understand the physical implication of a chiral symmetry. However, in models where a sublattice structure is apparent, we can associate chiral symmetry with the sublattice symmetry. This means that if the two sublattices of the system are interchanged, the physics remains invariant.

3A.2 Mathematical representation of the symmetries

We consider a Hamiltonian $\mathcal{H}(k)$ in the k-space. Let us recall what various symmetries do to the Hamiltonian, $\mathcal{H}$. The time reversal symmetry implies that $\mathcal{T}\mathcal{H}(k)\mathcal{T}^{-1} = \mathcal{H}(-k)$. For a particle–hole symmetric system, $\mathcal{C}\mathcal{H}(k)\mathcal{C}^{-1} = -\mathcal{H}(-k)$. Lastly, for a chiral symmetric system $\mathcal{S}\mathcal{H}(k)\mathcal{S}^{-1} = -\mathcal{H}(k)$. To understand the physical picture behind time reversal, it could be interesting to study its effect on the time evolution operator $e^{i\mathcal{H}t}$, $\mathcal{H}$ being the Hamiltonian for the system. We know that time reversal has the effect of complex conjugation, on spinless systems, that is,

$$\mathcal{T}i\mathcal{T}^{-1} = -i. \tag{3.5.2.1}$$

Hence, the time evolution operator looks like

$$\mathcal{T}e^{i\mathcal{H}t}\mathcal{T}^{-1} = e^{-i\mathcal{H}t} = e^{i\mathcal{H}(-t)}. \tag{3.5.2.2}$$

From this expression, it is apparent why time reversal operator has the name that it does. It changes t to $-t$ that results in the particle moving in the opposite direction. Another way to study the symmetry operators could be to talk in the language of second quantization, or more specifically, the creation–annihilation operators. Charge-conjugation causes a creation operator to transform into a superposition of annihilation operators, and vice versa. Time reversal on the other hand only has the effect of complex conjugation of the coefficients.

$$
\begin{aligned}
\mathcal{T}\psi_A\mathcal{T}^{-1} &= \sum_B (U_{\mathcal{T}}^*)_{A,B}\psi_B \\
\mathcal{T}\psi_A^\dagger\mathcal{T}^{-1} &= \sum_B (U_{\mathcal{T}})_{B,A}\psi_B \\
\mathcal{C}\psi_A\mathcal{C}^{-1} &= \sum_B (U_{\mathcal{C}}^*)_{A,B}\psi_B^\dagger \\
\mathcal{C}\psi_A^\dagger\mathcal{C}^{-1} &= \sum_B (U_{\mathcal{C}})_{B,A}\psi_B.
\end{aligned}
\tag{3.5.2.3}
$$

TABLE 3A.1. Table for tenfold symmetry classification. Different symmetry classes are indicated in the bottom row.

$\mathcal{T}$	1	1	1	-1	-1	-1	0	0	0	0
$\mathcal{C}$	1	-1	0	1	-1	0	1	-1	0	0
$\mathcal{S} = \mathcal{TC}$	1	1	0	1	1	0	0	0	0	1
	BDI	CI	AI	BIII	CII	AII	D	C	A	AIII

The classes are differentiated on the basis of whether the time reversal and charge-conjugation operators square to 1, -1 or 0. This gives a total of 9 separate classes. However, as previously discussed, there is another class having only the chiral symmetry. This corresponds to purely chiral systems and constitutes the tenth class in the classification scheme. Two broad sub-groups are apparent from this table. The first one is the purely chiral AIII group which only hosts chiral systems. The second one is the D group which only has charge-conjugation symmetry. This group hosts systems that are superconducting. Also, there are classes like BDI which hosts both chiral and superconducting systems.

Next, we venture upon understanding how this classification originated and wherein lies its importance. The idea behind the tenfold classification is entirely mathematical and was introduced by Elie Cartan, way back in the 1920s. Cartan's classification of $N \times N$ Hermitian matrices into 10 different groups bears one-to-one correspondence with this tenfold fermionic symmetry classification. The form of classification that we use today was introduced by Altland and Zirnbauer in 1997 [40]. The importance of this classification lies in the fact that corresponding to every class and pertaining to the dimension of our system, we may or may not have a distinct topological invariant. This invariant is a signature of the bulk–boundary correspondence and hints towards the presence of edge characteristics that make our system topologically non-trivial, and hence interesting.

The $\mathbb{Z}$-type invariant refers to an integer classification, whereas the $\mathbb{Z}_2$ invariant refers to a binary classification. A zero (0) at any position in the table refers to the fact that no topological insulator or superconductor can be found corresponding to that class and dimension. It would be interesting to associate a few known examples with the given table of invariants. A 2D system belonging to the class 'A' resembles our well-known integer quantum Hall system, thereby making the Chern number to be a $\mathbb{Z}$ invariant. From the table, we find that it has a $\mathbb{Z}$-type topological invariant. This, as we know, is indeed true and the $\mathbb{Z}$ invariant can be

TABLE 3A.2. Table for the topological invariants in different symmetry classes.

AZ/d	1	2	3
A	0	$\mathbb{Z}$	0
AIII	$\mathbb{Z}$	0	$\mathbb{Z}$
AI	0	0	0
BDI	$\mathbb{Z}$	0	0
D	$\mathbb{Z}_2$	$\mathbb{Z}$	0
DIII	$\mathbb{Z}_2$	$\mathbb{Z}_2$	$\mathbb{Z}$
AII	0	$\mathbb{Z}_2$	$\mathbb{Z}_2$
CII	$\mathbb{Z}$	0	$\mathbb{Z}_2$
C	0	$\mathbb{Z}$	0
CI	0	0	$\mathbb{Z}$

TABLE 3A.3. Periodic table of topological insulators and superconductors.

AZ/d	0	1	2	3	4	5	6	7	8	9
A	$\mathbb{Z}$	0	$\mathbb{Z}$	0	$\mathbb{Z}$	0	$\mathbb{Z}$	0	$\mathbb{Z}$	...
AIII	0	$\mathbb{Z}$	0	$\mathbb{Z}$	0	$\mathbb{Z}$	0	$\mathbb{Z}$	0	...
AI	$\mathbb{Z}$	0	0	0	$\mathbb{Z}$	0	$\mathbb{Z}_2$	$\mathbb{Z}_2$	$\mathbb{Z}$	...
BDI	$\mathbb{Z}_2$	$\mathbb{Z}$	0	0	0	$\mathbb{Z}$	0	$\mathbb{Z}_2$	$\mathbb{Z}_2$	...
D	$\mathbb{Z}_2$	$\mathbb{Z}_2$	$\mathbb{Z}$	0	0	0	$\mathbb{Z}$	0	$\mathbb{Z}_2$	...
DIII	0	$\mathbb{Z}_2$	$\mathbb{Z}_2$	$\mathbb{Z}$	0	0	0	$\mathbb{Z}$	0	...
AII	$\mathbb{Z}$	0	$\mathbb{Z}_2$	$\mathbb{Z}_2$	$\mathbb{Z}$	0	0	0	$\mathbb{Z}$	...
CII	0	$\mathbb{Z}$	0	$\mathbb{Z}_2$	$\mathbb{Z}_2$	$\mathbb{Z}$	0	0	0	...
C	0	0	$\mathbb{Z}$	0	$\mathbb{Z}_2$	$\mathbb{Z}_2$	$\mathbb{Z}$	0	0	...
CI	0	0	0	$\mathbb{Z}$	0	$\mathbb{Z}_2$	$\mathbb{Z}_2$	$\mathbb{Z}$	0	...

directly associated with the conductivity of the system. Another example would be the quantum spin Hall system, which belongs to the class AII in 2D. It is characterized by a $\mathbb{Z}_2$ topological invariant. Similarly, a 1D SSH chain belongs to the BDI class. A 1D Kitaev chain of spinless particles with p-wave superconducting correlations obey time reversal and particle–hole (charge-conjugation) symmetries and hence belong to the BDI class with $\mathbb{Z}$ as the topological invariant. However, if one talks about spinor particles, that is, electrons and an external magnetic field is included in the discussion, the time reversal symmetry vanishes. This places the Kitaev model in the D class (see Table 3A.3) which is characterized by the $\mathbb{Z}_2$ invariant.

The tenfold classification of symmetry shows certain interesting trends when studied carefully. In Table 3A.3, we find that corresponding to every dimension there are only 5 classes which host topologically non-trivial materials. Furthermore, for every class, there exists a periodicity of 8. The invariants repeat themselves after every 8th dimension. Additionally, we find that if the topological invariant for a particular class in the nth dimension is known, we know the invariant for the $(n + 1)^{th}$ dimension in the successive class. The origin of the above periodicities can be understood from a discussion on group theory.

4

Quantum Hall Effect in Graphene

4.1 Introduction

Graphene is formed of C atoms. C is an element in the IVth column of the periodic table and has four valence electrons in the outermost shell. It can make two types of chemical bonds, namely sp^3, which results in diamond known from ancient times, and a more stable sp^2, which results in graphite that is known for the last 500 years. A quick look at the discoveries of different allotropes of C is available in Table 4.1. The sp^2 hybridization causes planar configuration involving 3 of the 4 electrons, which are $120°$ apart and are bound by σ bonds that add stiffness (and flatness too) to the linkage between the C–C atoms, while the fourth electron bound to the C atoms via the π bond projects out of the plane, and is available for conduction. Thus, the electronic structure that we shall be discussing elaborately is due to these π electrons.

Graphene was the first discovery of atomically thin perfect two-dimensional (2D) material. Andre Geim and co-workers successfully exfoliated graphene from graphite [2, 3]. Some of the remarkable properties of graphene (which, unfortunately, we shall not worry too much about) include its strength, impermeability, very large thermal conductivity (at least one order larger than copper), as a molecule sensor, transparent (for its usage in displays), in the field of biology, such as neuron growth and DNA sequencing, and many more. Owing to the tremendous fundamental and technological applications of graphene, the discovery earned a Nobel Prize to A. Geim and K. Novoselov, both from the University of Manchester in the UK in 2010.

TABLE 4.1. The chronological discoveries of different allotropes of carbon.

Year/period of discovery	C allotropes
4000 BC	Diamond (3D)
16th century	Graphite (3D)
1985	C^{60}, Buckyball (0D)
1991	Multi-wall Carbon (1D) Nanotube (MWCNT)
1993	Single-wall Carbon (1D) Nanotube (SWCNT)
2005	Graphene

4.2 Tight-binding Hamiltonian

Having studied the quantized Hall effect in a 2DEG in detail, we focus on another system that is of topical interest, namely graphene. Apart from reviewing the basic electronic properties of graphene, we discuss the properties of the Landau levels. The unequal spacing between the successive Landau levels is a feature that shows up in many experiments. However, before we embark on the Hall effect in graphene, let us review its electronic properties. Even prior to that let us have a brief discussion on the tight-binding models in general.

It is fairly well known that the free electron approximation is grossly inadequate to explain the physical properties of materials. The simplest modification to the free electron theory in terms of inclusion of the lattice effects could explain a large number of features. For example, the Bragg scattering of the electron waves on a lattice opens up spectral gaps in the available energy levels at the boundaries of the first Brillouin zone (BZ), which are adequate in explaining the distinction between metals, semiconductors and insulators. Assuming that we have enough reasons for neglecting the electron–electron interaction, inclusion of only the electron–ion interaction in a crystalline solid often provides a good starting point.

The picture that emerges for an electron in a crystalline solid is that it sees a periodic potential due to atoms or ions placed at regular intervals as shown in Fig. 4.1, that is,

$$V(\mathbf{r}) = V(\mathbf{r} + \mathbf{R}),$$

where $\mathbf{R}$ denotes the periodicity of the lattice. The solution to the Schrödinger equation in the presence of such a potential is given by Bloch's theorem, which

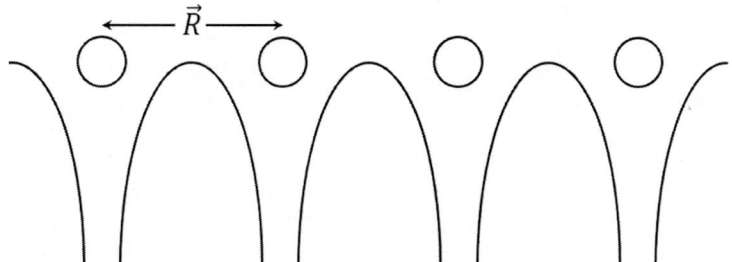

FIGURE 4.1. A schematic plot showing a periodic potential where each of the ionic locations are separated by **R**.

is stated as

$$\psi_{n,\mathbf{k}}(\mathbf{r}) = e^{i\mathbf{k}\cdot\mathbf{r}} u_{n,\mathbf{k}}(\mathbf{r}), \tag{4.2.1}$$

where the envelope function, $u_{n,\mathbf{k}}(\mathbf{r})$, picks up the periodicity of the lattice, that is,

$$u_{n,\mathbf{k}}(\mathbf{R} + \mathbf{r}) = u_{n,\mathbf{k}}(\mathbf{r}),$$

where **k** is the wavevector and n denotes the band index.

There is more than one way of proving Bloch's theorem. We shall sketch a simple proof using the translation operator, $T(\mathbf{R})$, which when acts on a wavefunction, $\psi(\mathbf{r})$, translates the wavefunction to $\mathbf{r} + \mathbf{R}$, that is,

$$T(\mathbf{R})\psi(\mathbf{r}) = \psi(\mathbf{r} + \mathbf{R}).$$

R connects one lattice point to another. Applying it to a generic Hamiltonian, $\mathcal{H}(\mathbf{p}, \mathbf{r}) = \frac{\mathbf{p}^2}{2m} + V(\mathbf{r})$ that obeys $\mathcal{H}\psi(\mathbf{r}) = E\psi(\mathbf{r})$, one gets

$$T(\mathbf{R})[\mathcal{H}(\mathbf{p},\mathbf{r})\psi(\mathbf{r})] = T(\mathbf{R})[E\psi(\mathbf{r})], \tag{4.2.2}$$

It implies that[1]

$$\mathcal{H}(\mathbf{p},\mathbf{r}+\mathbf{R})\psi(\mathbf{r}+\mathbf{R}) = E\psi(\mathbf{r}+\mathbf{R})]. \tag{4.2.3}$$

Due to the presence of the periodic potential, $V(\mathbf{r}) = V(\mathbf{r} + \mathbf{R})$,

$$\mathcal{H}(\mathbf{p} + \mathbf{r} + \mathbf{R}) = \mathcal{H}(\mathbf{r}).$$

---

[1] $T(\mathbf{R})$ translates everything on its right by **R**.

Thus,

$$\mathcal{H}(\mathbf{p,r})\psi(\mathbf{r+R}) = E\psi(\mathbf{r+R}). \tag{4.2.4}$$

Hence,

$$\mathcal{H}(\mathbf{p,r})T(\mathbf{R})\psi(\mathbf{r}) = ET(\mathbf{R})\psi(\mathbf{r}). \tag{4.2.5}$$

Putting it together

$$[\mathcal{H}, T] = 0. \tag{4.2.6}$$

Since $\mathcal{H}$ and T commute, they must share a common eigenfunction. Let $\psi(\mathbf{r})$ is the eigenfunction, such that

$$T(\mathbf{R})\psi(\mathbf{r}) = \psi(\mathbf{r+R}) = \lambda(\mathbf{R})\psi(\mathbf{r}), \tag{4.2.7}$$

$\lambda(\mathbf{R})$ being the corresponding eigenvalue. Applying the translation operator twice, one gets

$$T^2(\mathbf{R})\psi(\mathbf{r}) = \psi(\mathbf{r+2R}) = \lambda^2(\mathbf{R})\psi(\mathbf{r}). \tag{4.2.8}$$

Similarly, operating $T(\mathbf{R})$ N times yields

$$T^N(\mathbf{R})\psi(\mathbf{r}) = \psi(\mathbf{r+NR}) = \lambda^N(\mathbf{R})\psi(\mathbf{r}). \tag{4.2.9}$$

Now, if N denotes the number of unit cells, a periodic boundary condition will imply

$$\psi(\mathbf{r+NR}) = \psi(\mathbf{r}),$$

which demands

$$\lambda^N(\mathbf{R}) = 1. \tag{4.2.10}$$

A solution of the above equation has to be of the form

$$\lambda(\mathbf{R}) = e^{i\mathbf{k.R}}, \tag{4.2.11}$$

with $\mathbf{k} = \frac{m\mathbf{a}^*}{N}$, $\mathbf{a}^*$ being the reciprocal lattice vectors and m is an integer. Thus, we arrive at Bloch's theorem,

$$\psi(\mathbf{r+R}) = e^{i\mathbf{k.R}}\psi(\mathbf{r}). \tag{4.2.12}$$

A few comments are in order. They are as follows:

1. We have not specified the strength of $V(\mathbf{r})$, hence Bloch's theorem is valid for arbitrarily large electron–ion interactions, and the travelling wave description of the electrons still holds.
2. While the Hamiltonian is invariant under the translation caused by $T(\mathbf{R})$, the wavefunction is not and has an additional phase factor.
3. The $\mathbf{k}$ vectors are restricted to the first BZ, since for any arbitrary $\mathbf{k}'$ outside the first BZ, it can always be brought back to the first BZ using $\mathbf{k}' = \mathbf{k} + \mathbf{G}$, with $\mathbf{k}$ belonging to the first BZ and $\mathbf{G}$ is a reciprocal lattice vector.
4. The above makes the energy level to be periodic as well, namely

$$E(\mathbf{k}) = E(\mathbf{k} + \mathbf{G}).$$

5. $\mathbf{k}$ is *not* the momentum, and has a special name called *crystal momentum*, which can be checked via operating the momentum operator ($\mathbf{p} = -i\hbar\nabla$) on $\psi_{n\mathbf{k}}$, which yields

$$\mathbf{p} = \hbar\mathbf{k} + \mathcal{O}(\nabla u_{n,\mathbf{k}}).$$

Thus, $\mathbf{k}$ can be interpreted as a quantum number that describes a Bloch state.
6. The electron velocities are given by

$$\mathbf{v}_{n,\mathbf{k}} = \frac{1}{\hbar}\nabla_{\mathbf{k}}\epsilon_{n,\mathbf{k}},$$

where $\mathbf{v}_{n,\mathbf{k}}$ denotes the group velocity (v_g), $\nabla_{\mathbf{k}}$ is the gradient operator acting in $\mathbf{k}$-space, and $\epsilon_{n,\mathbf{k}}$ denotes the band energies, which are yet to be ascertained.

A very useful quantity in this regard is the *effective mass*, which parameterizes the dynamics of the Bloch electrons when they are subjected to an external force; for example, the force experienced by a charge particle in an electric field. In the presence of such a force field, $\mathbf{F}$, the work done or the change in energy within a time, δt, is given by

$$\delta E = \mathbf{F} \cdot \mathbf{v}\delta t. \tag{4.2.13}$$

Equivalently,

$$\frac{d\epsilon_{\mathbf{k}}}{dk}\delta k = \hbar v d\mathbf{k}. \tag{4.2.14}$$

Equating the above two equations (written in scalar form) yields

$$|\mathbf{F}| = \hbar \frac{d|\mathbf{k}|}{dt}. \tag{4.2.15}$$

Now taking a time derivative of the velocity expression

$$\frac{dv}{dt} = \frac{1}{\hbar} \frac{d}{dt} \left(\frac{d\epsilon_k}{dk} \right) = \frac{1}{\hbar} \left(\frac{d^2 \epsilon_k}{dk^2} \right) \frac{dk}{dt}. \tag{4.2.16}$$

Substituting Eq. (4.2.15) in the above equation

$$\frac{dv}{dt} = \frac{1}{\hbar^2} \left(\frac{d^2 \epsilon_k}{dk^2} \right) |\mathbf{F}|. \tag{4.2.17}$$

Using

$$F = m^* \frac{dv}{dt}$$

with m^* denoting the effective mass and is given by

$$m^* = \hbar^2 \left(\frac{\partial^2 \epsilon_k}{\partial k^2} \right)^{-1}. \tag{4.2.18}$$

For a free particle whose energy dispersion is given by $\epsilon_k = \frac{\hbar^2 k^2}{2m}$, one gets $m^* = m$, where m denotes the bare mass of the particle.

In general, the effective mass is a direction-dependent quantity and hence is a tensor. It is in fact a symmetric tensor of rank 2 and is denoted by m^*_{ij} which depends upon the *curvature* of the energy dispersion via

$$m^*_{ij} = \hbar^2 \left(\frac{\partial^2 \epsilon_k}{\partial k_i \partial k_j} \right)^{-1}. \tag{4.2.19}$$

Further, m^* can either be positive or negative depending on the sign of the curvature of the band dispersion, which is in sharp contrast with the notion of the bare mass. Also for an isotropic and parabolic dispersion in three-dimensional (3D), the density of states is given by

$$g(\epsilon) \simeq \left(\frac{2m^*}{\hbar^2} \right) \tag{4.2.20}$$

up to certain constants. Thus, a large effective mass (caused by large curvature of the energy dispersion) implies a larger density of states.

Finally, in order to find $\epsilon_{n,\mathbf{k}}$, we have to resort to an approximation. There are several approximate methods to compute $\epsilon_{n,\mathbf{k}}$ in real materials; however, a simple yet impactful method is the *tight-binding approximation*, which we briefly outline below. The essence of the approximation is that the electrons are tightly bound to the ionic sites, and hence the electronic states are mostly peaked at those sites. Although there is a small overlap between the wavefunctions localized at the neighbouring sites, just enough to impart kinetic energy to the electrons.

Consider a generic Hamiltonian of the form

$$\mathcal{H} = \frac{p^2}{2m} + \sum_i V_i = K + \sum_i V_i, \tag{4.2.21}$$

where the first term denotes the kinetic energy of the electrons and the second term is the periodic ionic potential that the electron is subjected to (there is no electron–electron interaction here). We can make a reasonable ansatz for the solution to be of the form

$$|\psi_{\mathbf{k}}\rangle = \sum_\alpha c_{\mathbf{k}} |\phi_\alpha\rangle. \tag{4.2.22}$$

$|\psi_{\mathbf{k}}\rangle$ obeys the Bloch theorem and $|\phi_\alpha\rangle$ denotes the basis states. Now we can diagonalize $\mathcal{H}$ and the matrix elements of $\mathcal{H}$ can be written as

$$\begin{aligned}
\mathcal{H}_{\alpha\beta} &= \langle\phi_\alpha|\mathcal{H}|\phi_\beta\rangle \tag{4.2.23} \\
&= \langle\phi_\alpha|(K + \sum_i V_i)|\phi_\beta\rangle \\
&= \epsilon_{\text{at}} + \langle\phi_\alpha|\sum_i V_i|\phi_\beta\rangle,
\end{aligned}$$

where ϵ_{at} denote the onsite atomic energies.

To reduce the complexity, let us specialize to the case of a 1D chain, so that the following simplification is achieved. Making an ansatz of the form

$$\begin{aligned}
\langle\phi_\alpha|\sum_i V_i|\phi_\beta\rangle &= V_0 \quad \text{for } \alpha = \beta \tag{4.2.24} \\
&= -t \quad \text{for } \alpha = \beta \pm 1 \\
&= 0 \quad \text{otherwise,}
\end{aligned}$$

where t denotes the strength of the coupling between the neighbouring sites of α. In this approximation,

$$\mathcal{H}_{\alpha\beta} = \epsilon_0 \delta_{\alpha,\beta} - t(\delta_{\alpha+1,\beta} + \delta_{\alpha-1,\beta}), \tag{4.2.25}$$

with $\epsilon_0 = \epsilon_{at} + V_0$. Finally, using the Bloch states, the matrix elements can be written as

$$\langle \psi_{\mathbf{k}} | \mathcal{H} | \psi_{\mathbf{k}} \rangle = \sum_{\alpha\beta} e^{-i\mathbf{k}.\mathbf{r}_\alpha} \langle \phi_\alpha | \mathcal{H} | \phi_\beta \rangle e^{i\mathbf{k}.\mathbf{r}_\beta} \tag{4.2.26}$$

$$= \sum_\alpha \left[\epsilon_0 - t(e^{ika} + e^{-ika}) \right]$$

$$= N(\epsilon_0 - 2t \cos ka),$$

where N denotes the total number of sites, and a is the lattice constant, which is the distance between the neighbouring atoms or ions. Thus, the electronic dispersion can be written as

$$\epsilon_k = \epsilon_0 - 2t \cos ka. \tag{4.2.27}$$

Ignoring the onsite energy (ϵ_0) for the moment, and extending it to more than 1D for a one-band model, one can write the tight-binding dispersion as

$$\epsilon_{\mathbf{k}} = -2t \cos \mathbf{k}.\delta_i, \tag{4.2.28}$$

where δ_i connects to the neighbouring sites. For a 2D square lattice with $\delta_i = \pm a\hat{x} \pm a\hat{y}$, the tight-binding dispersion is given by

$$\epsilon_{\mathbf{k}} = -2t(\cos k_x a + \cos k_y a). \tag{4.2.29}$$

We show the surface plot of the corresponding energy dispersion in the first BZ $(-\frac{\pi}{a} \leq k_{x,y} \leq +\frac{\pi}{a})$ (Fig. 4.2). The constant energy contours are visible for different electron fillings.

In the following section, we shall extend these ideas for a two-band model in 2D, namely graphene.

4.2.1   Basic electronic properties of graphene

Owing to the large mobility of π electrons in graphene, a nearest neighbour tight-binding Hamiltonian of the following form is most suitable, namely

$$\mathcal{H} = -t \sum_{\langle ij \rangle, \sigma} (a_{i\sigma}^\dagger b_{j\sigma} + \text{h.c.}), \tag{4.2.1.1}$$

where $a^\dagger(b)$ are the creation (annihilation) operators for electrons corresponding to sublattice A(B). Here σ represents the spin of the electrons, which is suppressed in

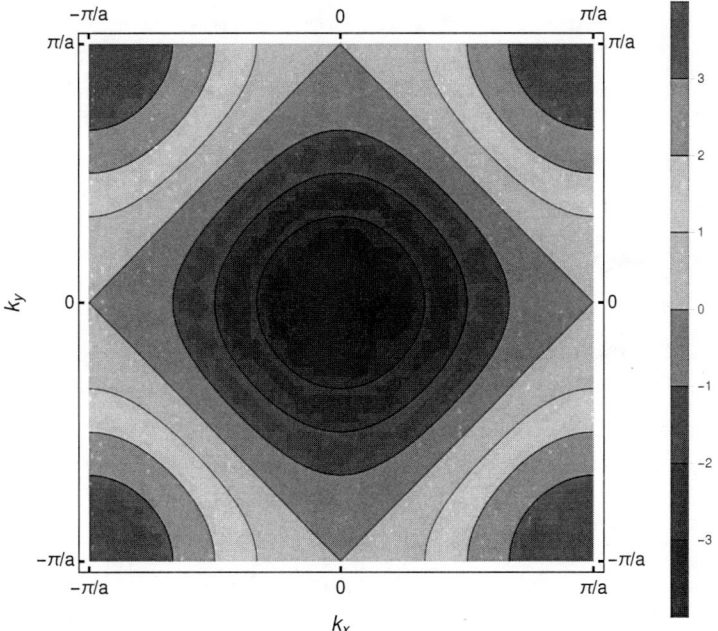

FIGURE 4.2. A surface plot of the energy dispersion on a 2D square lattice is shown.

the ongoing calculation because it plays no active role in the band structure or in the quantum Hall effect (QHE). We shall make the spin degrees of freedom apparent only when it is needed. Further, the nearest neighbour hopping t has value nearly 2.7eV, which is considerably large and we may ignore the electron–electron interaction. The vectors connecting the nearest neighbours, direct lattice vectors $\mathbf{a}_i$ and the reciprocal lattice vectors $\mathbf{b}_i$ (see Fig. 4.3) are written as

$$\boldsymbol{\delta}_1 = \frac{a}{2}\left(\sqrt{3}\hat{x} + \hat{y}\right); \quad \boldsymbol{\delta}_2 = \frac{a}{2}\left(-\sqrt{3}\hat{x} + \hat{y}\right); \quad \boldsymbol{\delta}_3 = -a\hat{x} \qquad (4.2.1.2)$$

$$\mathbf{a}_1 = \frac{a}{2}\left(\sqrt{3}\hat{x} + 3\hat{y}\right); \quad \mathbf{a}_2 = \frac{a}{2}\left(-\sqrt{3}\hat{x} + 3\hat{y}\right)$$

$$\mathbf{b}_1 = \frac{2\pi}{3a}\left(\sqrt{3}\hat{k}_x + \hat{k}_y\right); \quad \mathbf{b}_2 = \frac{2\pi}{3a}\left(-\sqrt{3}\hat{k}_x + 3\hat{k}_y\right),$$

where $a = 2.46\mathring{A}$ is the lattice constant. We define the lattice vector $\mathbf{R}$ at an arbitrary site as

$$\mathbf{R} = n\mathbf{a}_1 + m\mathbf{a}_2 \quad n, m \in N. \qquad (4.2.1.3)$$

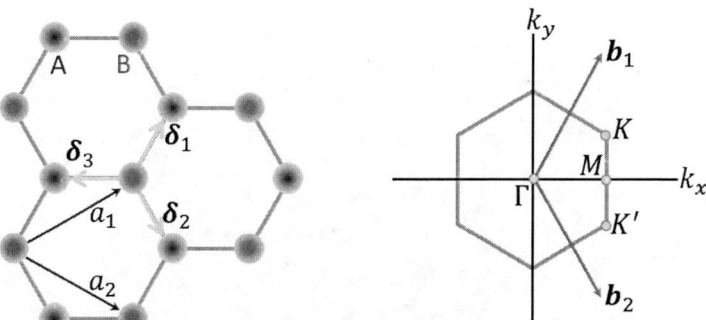

FIGURE 4.3. Plot showing nearest neighbour vectors δ_i, direct lattice vectors $\mathbf{a}_i$ (left-hand panel) and the reciprocal lattice vectors $\mathbf{b}_i$ (right-hand panel) for graphene. The circles denote the C atoms on A and B sublattices in the left-hand panel, while the first BZ is depicted in the right-hand panel.

Now, using the lattice vector, $\mathbf{R}$, and the nearest neighbour vectors, δ_i, we explicitly write the tight-binding Hamiltonian as

$$\mathcal{H} = -t \sum_{\mathbf{R},\delta} \left[b^\dagger(\mathbf{R} + \delta_i) a(\mathbf{R}) + a^\dagger(\mathbf{R}) b(\mathbf{R} + \delta_i) \right] . \qquad (4.2.1.4)$$

In order to write the Hamiltonian in the momentum space, we Fourier transform the creation and annihilation operators using

$$a_\mathbf{k} = \frac{1}{\sqrt{N}} \sum_\mathbf{R} e^{-i\mathbf{k}\cdot\mathbf{R}} a(\mathbf{R}), \qquad (4.2.1.5)$$

which yields the Hamiltonian in the following form

$$\mathcal{H} = -\frac{t}{N} \sum_{\mathbf{k},\mathbf{q}} \sum_\mathbf{R} \sum_{i=1}^{3} e^{i(\mathbf{k}-\mathbf{q})\cdot\mathbf{R}} \left[e^{i(\mathbf{q}-\mathbf{k})\cdot\mathbf{R}} e^{i\mathbf{q}\cdot\delta_i} b_\mathbf{q}^\dagger a_\mathbf{k} + e^{i(\mathbf{k}-\mathbf{q})\cdot\mathbf{R}} e^{-i\mathbf{q}\cdot\delta_i} a_\mathbf{q}^\dagger b_\mathbf{k} \right] . \qquad (4.2.1.6)$$

The following definition of the Kronecker delta can be used

$$\delta_{\mathbf{k},\mathbf{q}} = \frac{1}{N} \sum_\mathbf{R} e^{i(\mathbf{k}-\mathbf{q})\cdot\mathbf{R}} \qquad (4.2.1.7)$$

to rewrite the Hamiltonian as

$$\mathcal{H} = -t \sum_{\mathbf{k}} \sum_{i=1}^{3} \left[e^{-i\mathbf{k}\cdot\boldsymbol{\delta}_i} b_{\mathbf{k}}^\dagger a_{\mathbf{k}} + e^{i\mathbf{k}\cdot\boldsymbol{\delta}_i} a_{\mathbf{k}}^\dagger b_{\mathbf{k}} \right] \tag{4.2.1.8}$$

$$= -t \sum_{\mathbf{k}} \sum_{i=1}^{3} \begin{pmatrix} a_{\mathbf{k}}^\dagger & b_{\mathbf{k}}^\dagger \end{pmatrix} \begin{pmatrix} 0 & e^{-i\mathbf{k}\cdot\boldsymbol{\delta}_i} \\ e^{i\mathbf{k}\cdot\boldsymbol{\delta}_i} & 0 \end{pmatrix} \begin{pmatrix} a_{\mathbf{k}} \\ b_{\mathbf{k}} \end{pmatrix}$$

$$= -t \sum_{\mathbf{k}} \sum_{i=1}^{3} \begin{pmatrix} a_{\mathbf{k}}^\dagger & b_{\mathbf{k}}^\dagger \end{pmatrix} h(\mathbf{k}) \begin{pmatrix} a_{\mathbf{k}} \\ b_{\mathbf{k}} \end{pmatrix},$$

where $h(\mathbf{k})$ is the Hamiltonian matrix defined by

$$h(\mathbf{k}) = -t \begin{pmatrix} 0 & \left(e^{i\mathbf{k}\cdot\boldsymbol{\delta}_1} + e^{i\mathbf{k}\cdot\boldsymbol{\delta}_2} + e^{i\mathbf{k}\cdot\boldsymbol{\delta}_3} \right) \\ \left(e^{-i\mathbf{k}\cdot\boldsymbol{\delta}_1} + e^{-i\mathbf{k}\cdot\boldsymbol{\delta}_2} + e^{-i\mathbf{k}\cdot\boldsymbol{\delta}_3} \right) & 0 \end{pmatrix}. \tag{4.2.1.9}$$

Since the difference between two nearest neighbour lattice vectors, $\boldsymbol{\delta}_i$ and $\boldsymbol{\delta}_j$, must yield a lattice vector, $\mathbf{R}$, we can do a transformation,

$$a_{\mathbf{k}} \to e^{i\mathbf{k}\cdot\boldsymbol{\delta}_3} a_{\mathbf{k}} \quad \text{and} \quad a_{\mathbf{k}}^\dagger \to e^{-i\mathbf{k}\cdot\boldsymbol{\delta}_3} a_{\mathbf{k}}^\dagger.$$

The above form helps to write a new Hamiltonian matrix,

$$\tilde{h}(\mathbf{k}) = -t \begin{pmatrix} 0 & \left(e^{i\mathbf{k}\cdot(\boldsymbol{\delta}_1-\boldsymbol{\delta}_3)} + e^{i\mathbf{k}\cdot(\boldsymbol{\delta}_2-\boldsymbol{\delta}_3)} + 1 \right) \\ \left(e^{-i\mathbf{k}\cdot(\boldsymbol{\delta}_1-\boldsymbol{\delta}_3)} + e^{-i\mathbf{k}\cdot(\boldsymbol{\delta}_2-\boldsymbol{\delta}_3)} + 1 \right) & 0 \end{pmatrix}. \tag{4.2.1.10}$$

Using the definitions of $\boldsymbol{\delta}_i$, one gets

$$\tilde{h}(\mathbf{k}) = -t \begin{pmatrix} 0 & -\left(e^{i\mathbf{k}\cdot\mathbf{a}_1} + e^{i\mathbf{k}\cdot\mathbf{a}_2} + 1 \right) \\ -\left(e^{-i\mathbf{k}\cdot\mathbf{a}_1} + e^{-i\mathbf{k}\cdot\mathbf{a}_2} + 1 \right) & 0 \end{pmatrix}. \tag{4.2.1.11}$$

One may verify that $\tilde{h}(\mathbf{k})$ obeys $\tilde{h}(\mathbf{k}) = \tilde{h}(\mathbf{k}+\mathbf{G})$, where $\mathbf{G}$ is the reciprocal lattice vector, defined as $\mathbf{G} = p\mathbf{b}_1 + q\mathbf{b}_2$, with p and q being integers. Thus,

$$\tilde{h}(\mathbf{k}) = -t \begin{pmatrix} 0 & f(\mathbf{k}) \\ f^*(\mathbf{k}) & 0 \end{pmatrix}, \tag{4.2.1.12}$$

where

$$f(\mathbf{k}) = -t \left(e^{-ik_x a} + 2 e^{ik_x a/2} \cos\left(\frac{k_y \sqrt{3}a}{2} \right) \right).$$

The tight-binding energy is obtained by diagonalizing $\tilde{h}(\mathbf{k})$, which yields

$$\epsilon_{\mathbf{k}} = \pm t\sqrt{|f(k)|^2} = \pm t\sqrt{3 + 2\cos\left(\sqrt{3}ak_y\right) + 4\cos\left(\sqrt{3}ak_y/2\right)\cos(3ak_x/2)}.$$

(4.2.1.13)

The two bands described by the '+' and the '−' signs in the above dispersion touch at six points in the BZ (see Fig. 4.4). Since the C atom of graphene possesses a single accessible electron, one may assume a half-filled system where the lower band is completely filled. Further, we wish to discuss the low-lying excitations just above the ground state of the system.

This tells us to explore the low-energy theory of graphene. To achieve that, we have to obtain the points where the two bands touch each other in the BZ. Thus, we set $f(\mathbf{k}) = 0$, that is, we set both the real and the imaginary parts equal to zero as follows.

$$\cos(k_x a) + 2\cos(k_x a/2)\cos\left(\sqrt{3}k_y a/2\right) = 0 \qquad (4.2.1.14)$$

$$-\sin(k_x a) + 2\sin(k_x a/2)\cos\left(\sqrt{3}k_y a/2\right) = 0.$$

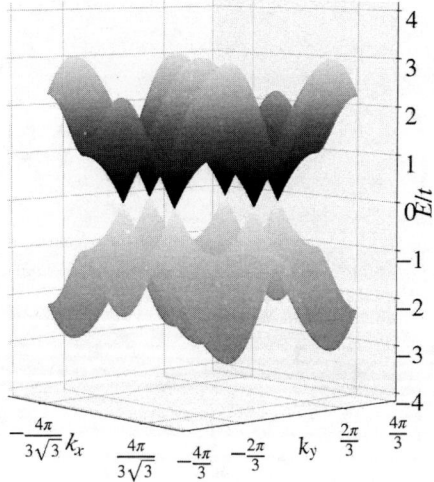

FIGURE 4.4. Plot showing the two tight-binding bands of graphene in the first BZ. The two bands touch at six points. In the vicinity of those points, the bands are linearly dispersing bearing signatures of (pseudo-)relativistic physics.

Eq. (4.2.1.14) can be manipulated as follows

$$\sin(k_x a/2)\left[-\cos(k_x a/2)+\cos\left(k_y a\sqrt{3}/2\right)\right]=0. \qquad (4.2.1.15)$$

Therefore, we are left with two options, namely

either (i) $\sin(k_x a/2)=0$; which means $\cos(k_x a/2)=\pm 1$;

or (ii) $\cos(k_x a/2)=\cos\left(\sqrt{3}k_y a/2\right)$.

Option (i) results in

$$1+2\cos\left(k_y\sqrt{3}a/2\right)=0,$$

which gives the points $\left(0,\pm\frac{4\pi}{3\sqrt{3}a}\right)$ (plus or minus the reciprocal lattice vector, **G**). Whereas, option (ii) can be written as

$$\cos\left(k_y a\sqrt{3}\right)+2\cos^2(k_y a\sqrt{3}/2)=0.$$

Thus, we get four more points, which are $\pm\frac{2\pi}{3a}\left(1,\frac{1}{\sqrt{3}}\right)$ and $\pm\frac{2\pi}{3a}\left(1,-\frac{1}{\sqrt{3}}\right)$ (again plus or minus the reciprocal lattice vector, **G**).

A closer introspection yields all the six points are not independent. For example, the set of vectors, namely $\left(0,\pm\frac{4\pi}{3\sqrt{3}a}\right)$, $\frac{2\pi}{3a}\left(1,-\frac{1}{\sqrt{3}}\right)$ and $\frac{2\pi}{3a}\left(-1,-\frac{1}{\sqrt{3}}\right)$, can be connected to each other via the combination of the reciprocal lattice vectors, $\mathbf{b}_1$ and $\mathbf{b}_2$. For example,

$$\left(0,\frac{4\pi}{3\sqrt{3}a}\right)+\mathbf{b}_2=\frac{2\pi}{3a}\left(1,-\frac{1}{\sqrt{3}}\right) \qquad (4.2.1.16)$$

$$\left(0,\frac{4\pi}{3\sqrt{3}a}\right)-\mathbf{b}_1=\frac{2\pi}{3a}\left(-1,-\frac{1}{\sqrt{3}}\right).$$

The same is true for the other vectors,

$$\left(0,\frac{4\pi}{3\sqrt{3}a}\right),\ \frac{2\pi}{3a}\left(-1,\frac{1}{\sqrt{3}}\right),\ \frac{2\pi}{3a}\left(1,\frac{1}{\sqrt{3}}\right).$$

Therefore, only two vectors are found to be independent. Traditionally, they are called **K** and **K′**, which are written as

$$\mathbf{K}=\frac{2\pi}{3a}\left(1,\frac{1}{\sqrt{3}}\right),\ \text{and}\ \mathbf{K'}=\frac{2\pi}{3a}\left(1,-\frac{1}{\sqrt{3}}\right).$$

Any other independent pair is also a valid choice for **K** and **K′**.

It should be noted that since the band gap vanishes at the **K** and **K'** points, there are two branches of low-energy excitations, namely one of them with momentum close to **K** and the other close to **K'**. Since $f(\mathbf{k})$ is zero at the $\mathbf{k} = \mathbf{K}$ point, we use the Taylor series expansion of $f(\mathbf{k})$ near **K** about $\mathbf{q} = \mathbf{k} - \mathbf{K} = 0$.

$$f'(\mathbf{q}) = \frac{\partial f(\mathbf{k})}{\partial k_x}\bigg|_{(k_x - K_x)}(k_x - K_x) + \frac{\partial f(\mathbf{k})}{\partial k_y}\bigg|_{(k_y - K_y)}(k_y - K_y)$$

$$= \frac{3at}{2}(q_x + iq_y).$$

This yields the energy spectrum

$$\epsilon_{\mathbf{K}}(\mathbf{q}) = \hbar v_F(q_x + iq_y), \tag{4.2.1.17}$$

where $v_F = \frac{3at}{2\hbar} \simeq 10^6 ms^{-1}$ is the Fermi velocity. Similarly, the spectrum around **K'** points has the following form

$$\epsilon_{\mathbf{K'}}(\mathbf{q}) = \hbar v_F(q_x - iq_y). \tag{4.2.1.18}$$

Combining these two energy expressions, one may write

$$\epsilon_{\mathbf{K},\mathbf{K'}} = \hbar v_F \mathbf{q} \cdot \boldsymbol{\sigma}, \tag{4.2.1.19}$$

where the Pauli matrix is denoted by $\boldsymbol{\sigma} = \sigma_x, \sigma_y$ and **q** represents the planar vector (q_x, q_y). The electrons close to the **K** and **K'** points are called massless Dirac fermions, as they obey the Dirac equation without the 'mass' term.[2] It may be noted that

$$\epsilon_{\mathbf{K'}}(\mathbf{q}) = \epsilon_{\mathbf{K}}^*(\mathbf{q}). \tag{4.2.1.20}$$

This equation tells us that (as will be seen later) the electrons at the **K** point have opposite helicity as compared to that of the **K'** points.

To sum up our preliminary discussion on graphene, the electronic energy is governed by the equation,

$$\epsilon(q) \pm v_F|q|. \tag{4.2.1.21}$$

As can be noticed, the eigenvalues depend only on the magnitude of the planar wave **q** irrespective of its direction in the 2D plane. Further, the Hamiltonian on

---

[2]The Dirac equation is written in conventional notations as $\mathcal{H} = c\boldsymbol{\alpha} \cdot \mathbf{p} + \beta mc^2$, where α and β are Hermitian operators, which do not operate on the space and time variables. In the case of graphene, the second term is absent.

a formal note represents that of a massless $s = 1/2$ particle, such as a neutrino; however, the velocity of such particles is lesser than 300 times to that of the light. Also, the handedness (or the helicity) feature of neutrinos is inbuilt, where the electrons behave left-handed and right-handed at the $\mathbf{K}$ and $\mathbf{K'}$ points similar to neutrino respectively.

The above discussion produces a well-known ambiguity for the effective mass (m^*) in the following sense. The definition in Eq. (4.2.18) yields a divergent effective mass, while the dispersion corresponds to that of a massless electron. A compromise is possible via an alternative definition of the effective mass, namely

$$m^* = \hbar^2 k \left(\frac{\partial \epsilon_k}{\partial k}\right)^{-1}. \qquad (4.2.1.22)$$

The motivation for this formula comes from semiclassical arguments. The momentum of the particle is given by

$$p = \hbar k = m^* v_g, \qquad (4.2.1.23)$$

where v_g denotes the group velocity $\left(\frac{1}{\hbar}\left(\frac{\partial \epsilon_k}{\partial k}\right)\right)$ defined earlier. Putting the expression for v_g in Eq. (4.2.1.23), one gets

$$p = \frac{m^*}{\hbar}\left(\frac{\partial \epsilon_k}{\partial k}\right), \qquad (4.2.1.24)$$

which justifies the definition written earlier (see Eq. 4.2.1.22), that is,

$$m^* = \hbar p \left(\frac{\partial \epsilon_k}{\partial k}\right)^{-1} = \hbar^2 k \left(\frac{\partial \epsilon_k}{\partial k}\right)^{-1}. \qquad (4.2.1.25)$$

It may be noted that the above expression correctly reproduces the mass of the non-relativistic free particles.

There is another *mass* that is also used in the context of electrons in graphene. It is called the *cyclotron mass*, m_c, defined via

$$m_c = \frac{1}{2\pi}\left[\frac{\partial A(\epsilon)}{\partial \epsilon}\right]_{\epsilon=\epsilon_F}, \qquad (4.2.1.26)$$

where $A(\epsilon)$ denotes the area on the Fermi sheet in k-space and ϵ_F is the Fermi energy with

$$A(\epsilon) = \pi q^2 = \frac{\pi \epsilon^2}{v_F^2},$$

where v_F denotes the Fermi velocity. Thus,

$$m_c = \frac{\epsilon_F}{v_F^2} = \frac{k_F}{v_F} = \frac{\sqrt{\pi n}}{v_F}, \tag{4.2.1.27}$$

where k_F is the Fermi wavevector and n denotes the electron density. So the cyclotron mass, m_c, is proportional to the density of electrons at the Fermi surface.

Further, the electron density can be obtained as follows. The number of carriers in the vicinity of the Dirac points is given by

$$N = A \int_0^\epsilon g(\epsilon')d\epsilon' = \frac{A\nu}{2\pi} \int_0^{q(\epsilon)} \frac{d\epsilon'}{dq}dq, \tag{4.2.1.28}$$

where A denotes the surface area of the region under consideration. ν takes care of the valley and the spin degeneracies. Owing to the linear relationship of the Dirac spectrum (see Eq. 4.2.1.21), the total number of electrons can be written as

$$N = \frac{A\nu}{2\pi} \int_0^{q(\epsilon)} q\,dq. \tag{4.2.1.29}$$

Comparing Eqs. (4.2.1.28) and (4.2.1.29), the DOS ($g(\epsilon)$) is obtained as

$$g(\epsilon) = \frac{\nu}{2\pi} \frac{q}{\partial \epsilon / \partial q}. \tag{4.2.1.30}$$

Using Eq. (4.2.1.21),

$$g(\epsilon) = \frac{\nu}{2\pi\hbar^2 v_F^2} |\epsilon|. \tag{4.2.1.31}$$

Thus, for $\epsilon \simeq 0$, the DOS vanishes linearly with energy. For *ordinary* electrons in 2D, the DOS is a constant, that is, independent of energy.

4.2.2   Experimental confirmation of the Dirac spectrum

When a beam of monochromatic photons with an energy larger than the work function of a particular material interacts with the constituent charges (electrons) by incidenting on the surface of the sample, the electrons absorb the photons and thus possess sufficient energy to escape from the sample. By measuring the energy and the momentum of the photoelectrons and using energy–momentum conservations laws, one can derive the properties of the electrons prior to them being incident on the surface and relate them with those getting scattered. Angle-resolved photoemission spectroscopy (ARPES) can be a direct probe to

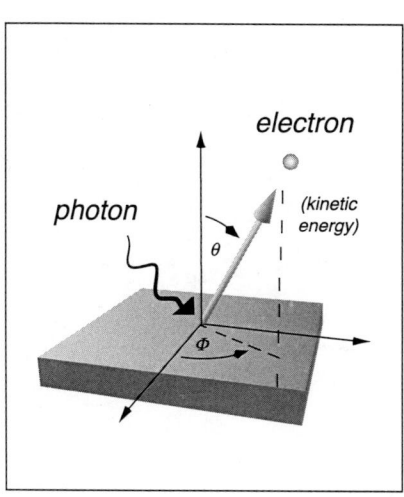

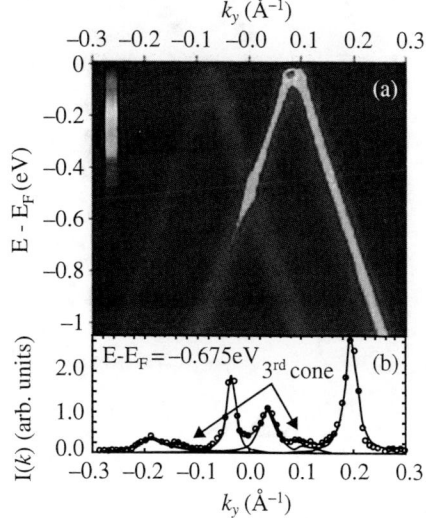

FIGURE 4.5. Plot showing the experimental setup for ARPES (left), the two linearly dispersing bands (middle) and the Dirac points are shown on the hexagonal BZ of graphene (right). Taken from Ref. [84].

resolve the momentum-dependent band structure and the topology of the Fermi surface. In ARPES, a photon is employed to eject an electron from the surface of the graphene layer. The intensity of the ARPES is proportional to the transition probability from an initial Bloch state with a crystal momentum, $\mathbf{k}$, and energy, E, to a final state, $\mathbf{k}'$. The method conclusively establishes the existence of Dirac fermions seen via linearly dispersing bands in the vicinity of the Dirac points. The experimental setup, ARPES data and the hexagonal BZ (which we have discussed before) are shown in Fig. 4.5.

4.3 Graphene nanoribbon

Before we proceed to explore the effects of magnetic field in graphene, a brief introduction to finite size systems is necessary. These are called nanoribbons. Graphene nanoribbons are elongated strips of graphene that may be obtained by cutting a long graphene sheet along a certain direction. Depending on the edge termination type, there can be two types of graphene nanoribbon, namely an armchair graphene nanoribbon (AGNR) and a zigzag graphene nanoribbon (ZGNR) as shown in Fig. 4.6. Along the x-direction shown in the figure, the GNR strips have periodic shapes. In the left figure, the GNR setup has an armchair shape, hence it is denoted by AGNR and the right one has a zigzag shape, thus it is called

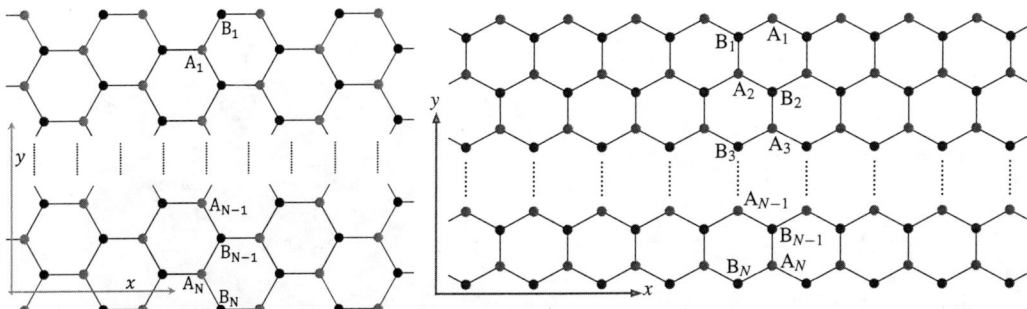

FIGURE 4.6. Schematic plot showing armchair (left) and zigzag (right) nanoribbons based on their edge termination type. Even though a schematic presentation of a zigzag ribbon is also presented elsewhere (see Fig. 4.7), it is included here once more for comparison with the armchair type.

ZGNR. It should also be noted that the width of the AGNR has a zigzag shape and likewise, a ZGNR has an armchair shape.

The electronic properties of GNRs depend on the geometry of the edges and lateral width of the nanoribbons. Wakabayashi *et al.* [85] examined analytically that the ZGNRs are always metallic, but the conducting state of the direct energy gap (ΔE) for AGNRs depend on their widths and follow the relations,

$$\Delta E \sim \begin{cases} 0 & N = 3M - 1 \\ -\frac{\pi}{W + \frac{\sqrt{3}}{2}} & N = 3M \\ \frac{\pi}{W} & N = 3M + 1 \end{cases}, \tag{4.3.1}$$

where W is the width of the AGNR and M is an integer. Hence, according to Eq. (4.3.1), AGNRs are always metallic when their width is $N = 3M - 1$, else they are semiconducting in nature.

In most of our numeric computation used in the book, we have used a ZGNR; however, the results for the AGNR will not be qualitatively different, except for the difference stated above.

4.3.1 Hofstadter butterfly

The fate of an electron gas or graphene described by a tight-binding model subjected to an external magnetic field is to show quantized plateaus of the Hall resistance. Consequently, the band energies of the electrons transform into discrete Landau levels. The presence of a periodic crystal potential adds further exciting features to the spectrum. The Hamiltonian of such a system is given by

$$\mathcal{H} = \frac{(\mathbf{p} - e\mathbf{A})^2}{2m} + V(\mathbf{r}), \tag{4.3.1.1}$$

where $V(\mathbf{r}) = V(\mathbf{r} + \mathbf{a})$ is the periodic potential with lattice periodicity $\mathbf{a}$. The electrons are described by the Bloch states leading to the formation of bands. In the presence of the vector potential, $\mathbf{A}$, each Bloch band gets further divided into sub-bands, and the resultant energy spectra as a function of the flux give rise to a fractal structure known as the Hofstadter butterfly. These rather complex energy spectra arise owing to a delicate interplay between the two length scales, namely a and l_B that are associated with two different quantization phenomena. In fact, the Hofstadter butterfly arises when the ratio of these two lengths is a rational fraction. An even more interesting scenario emerges when the ratio is not a rational fraction; however, we shall not discuss it here.

The fractal nature of the spectrum was observed by D. Hofstadter [86], which he obtained by solving Harper's equation [87], and demonstrated that for commensurate values of the magnetic flux such that $\Phi/\Phi_0 = p/q$, where the single particle Bloch bands split into q sub-bands, which themselves are p-fold degenerate (p and q being co-prime integers). Each of these p sub-bands further splits yielding a continued fraction as a function of the magnetic flux. The distance between the levels, sub-levels, and so on and the width of each of the 'superstructures' oscillates with variation of the magnetic field flux with a period that is universal and is independent of the form of the quasiparticle dispersion relation. Consequently, one observes a quasi-continuous distribution of incommensurate quantum states that form a self-similar recursive structure, like that of a butterfly. Schlösser *et al.* [88] have realized the Hofstadter spectrum for the first time in semiconductor superlattice structures. It was later observed for a number of systems, such as cold atomic systems in optical lattices. In continuation of our present discussion, we shall discuss the Hofstadter butterfly in graphene.

In order to demonstrate the Hofstadter butterfly [86], we have taken a semi-infinite ribbon of graphene. The ribbon geometry is such that it has zigzag edges as in Fig. 4.7. In tight-binding approximation, the external magnetic field enters through the hopping integral which is replaced by the Peierls substitution, namely

$$\exp\left(\frac{ie}{\hbar}\int_i^j \mathbf{A} \cdot d\mathbf{r}\right) t_{ij} = \exp\left(i(2\pi/\Phi_0)\int_i^j \mathbf{A} \cdot d\mathbf{r}\right) t_{ij}, \tag{4.3.1.2}$$

where t_{ij} is the hopping integral between the sites i and j with no field present. The flux is denoted in terms of flux quantum $\Phi_0 = h/e$. To include the magnetic field,

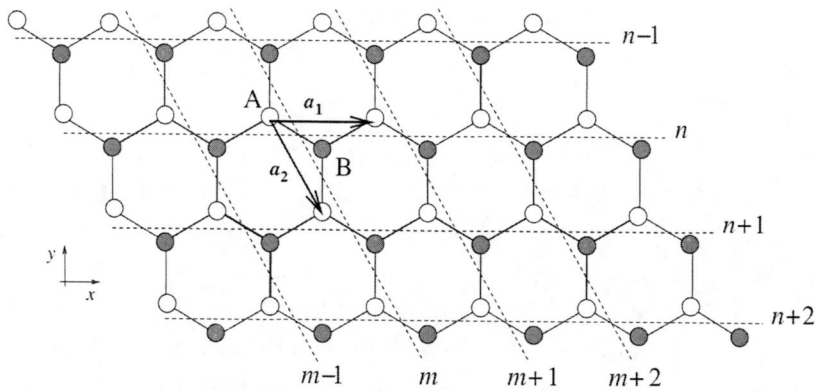

FIGURE 4.7. Geometry of a ZGNR is shown and is repeated to set up the notations. The open and filled circles represent A and B sublattices respectively. (m, n) denote the unit cell index in x and y directions.

we have taken $\mathbf{B} = B\hat{z}$ such that the vector potential ($\mathbf{A}$) in Landau gauge takes the form $\mathbf{A} = Bx\hat{y}$. With this modification, the tight-binding Hamiltonian for graphene introduced in Fig. 4.7 can be written as

$$\mathcal{H} = -\sum_{m,n} \left[t e^{i\pi \frac{\Phi}{\Phi_0} n} a^\dagger_{m,n} b_{m,n} + t e^{-i\pi \frac{\Phi}{\Phi_0} n} a^\dagger_{m,n} b_{m-1,n} + t_1 a^\dagger_{m,n} b_{m,n-1} + h.c. \right]. \quad (4.3.1.3)$$

Here $a^\dagger_{m,n}$, $b_{m,n}$ represent electron creation and annihilation operators of sublattice A and B respectively at site index (m, n). Since the ribbon is infinite in the x direction, one can use the Fourier decomposition of the operators for the index m giving the following equation [89]:

$$\mathcal{H} = -\sum_{k,n} \left[t e^{i\pi \frac{\Phi}{\Phi_0} n} a^\dagger_{k,n} b_{k,n} + t e^{-i\pi \frac{\Phi}{\Phi_0} n} e^{ika} a^\dagger_{k,n} b_{k,n} + t_1 a^\dagger_{k,n} b_{k,n-1} + h.c. \right]. \quad (4.3.1.4)$$

Assuming the eigenfunction to be $|\psi(k)\rangle = \sum_n [\alpha_{k,n} |a, k, n\rangle + \beta_{k,n} |b, k, n\rangle]$, the eigenvalue equation of Hamiltonian (4.3.1.4) gives the following two Harper [90] equations.

$$E_k \alpha_{k,n} = -\left[e^{ika/2} 2t \cos\left(\pi \frac{\Phi}{\Phi_0} n - \frac{ka}{2} \right) \beta_{k,n} + t_1 \beta_{k,n-1} \right] \quad (4.3.1.5)$$

$$E_k \beta_{k,n} = -\left[e^{-ika/2} 2t \cos\left(\pi \frac{\Phi}{\Phi_0} n - \frac{ka}{2} \right) \alpha_{k,n} + t_1 \alpha_{k,n+1} \right]. \quad (4.3.1.6)$$

In order to get the spectra for Haldane and semi-Dirac Haldane model, Eqs. (4.3.1.5) and (4.3.1.6) together with the following two characteristic equations of NNN hopping are used.

$$E_k \alpha_{k,n} = -t_2 [e^{ika/2} \cos\left(\frac{ka}{2} + \phi\right) \alpha_{k,n+1} \tag{4.3.1.7}$$

$$+ \cos(ka - \phi)\alpha_{k,n} + e^{-ika/2} \cos\left(\frac{ka}{2} + \phi\right) \alpha_{k,n-1}]$$

$$E_k \beta_{k,n} = -t_2 [e^{ika/2} \cos\left(\frac{ka}{2} - \phi\right) \beta_{k,n+1} \tag{4.3.1.8}$$

$$+ \cos(ka + \phi)\beta_{k,n} + e^{-ika/2} \cos\left(\frac{ka}{2} + \phi\right) \beta_{k,n-1}].$$

Using these equations we have numerically calculated the Hofstadter butterfly spectrum for the semi-infinite nanoribbon with $q = 200$. The fractal spectra are depicted in Fig. 4.8 for graphene [91]. Energies are taken in units of t. The fractal spectra result from the competition of magnetic field and lattice effects. For each effective flux $f = \phi/\phi_0 = p/q$, there are $2q$ number of sub-bands. The spectrum is periodic in f with periodicity $3\sqrt{3}Ba^2/2\phi_0$.

4.3.2   Landau levels in graphene

In order to proceed, we can write the Hamiltonian in a unified way that includes the description of both the Dirac points (valleys) **K** and **K'**. So for each valley, we have one 2D spinor Hamiltonian. Thus, augmenting the Hilbert space, we can write the eigenfunctions as

$$\psi = (\psi_{\mathbf{K'}}, \psi_{\mathbf{K}})^T$$

and the Hamiltonian is given by

$$\mathcal{H} = \hbar v_F \begin{pmatrix} -\boldsymbol{\sigma}^* \cdot \mathbf{k} & 0 \\ 0 & \boldsymbol{\sigma} \cdot \mathbf{k} \end{pmatrix} = v_F \begin{pmatrix} -\boldsymbol{\sigma}^* \cdot \mathbf{p} & 0 \\ 0 & \boldsymbol{\sigma} \cdot \mathbf{p} \end{pmatrix}. \tag{4.3.2.1}$$

Now we shall be discussing the motion of the massless relativistic electrons in a magnetic field. In a usual 2D electron gas, the Landau quantization produces equidistant levels (see Eq. 1.6.7), which is an artefact of the non-relativistic parabolic dispersion of the free carriers. We need to ascertain how the quantization formula is modified for the case of graphene.

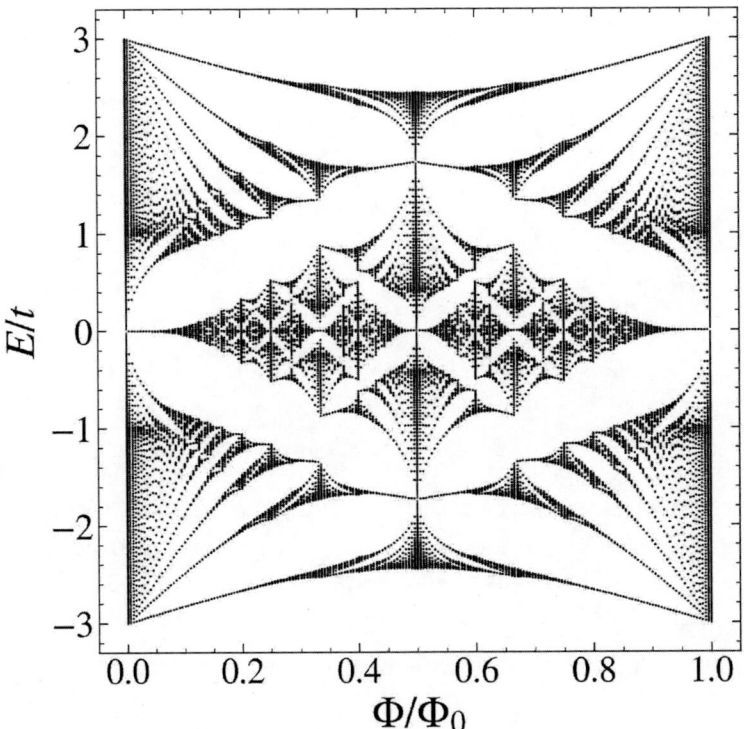

FIGURE 4.8. The Hofstadter butterfly is shown for graphene. The fractal structure as a function of the external flux (scaled by the flux quantum), that is, Φ/Φ_0 can be seen.

As earlier, we do the Peierls substitution, $\mathbf{p} \to \mathbf{p} + e\mathbf{A}$. Thus, the Hamiltonian becomes

$$\mathcal{H}_{\mathbf{K},\mathbf{K}'} = v_F \begin{pmatrix} 0 & -(p_x + ip_y) & 0 & 0 \\ -(p_x - ip_y) & 0 & 0 & 0 \\ 0 & 0 & 0 & (p_x - ip_y) \\ 0 & 0 & (p_x + ip_y) & 0 \end{pmatrix}.$$

The wavefunction has now four components, namely

$$\psi = \begin{pmatrix} \phi_A^{\mathbf{K}'} \\ \phi_B^{\mathbf{K}'} \\ \phi_A^{\mathbf{K}} \\ \phi_B^{\mathbf{K}} \end{pmatrix}, \tag{4.3.2.2}$$

where $\phi_{A,B}^{\mathbf{K}}$ are the wavefunctions for an electron at momentum values corresponding to the valley $\mathbf{K}$ at the two sublattice sites A and B. Similar notations carry on for the other valley $\mathbf{K}'$.

For a perpendicular magnetic field, $\mathbf{B} = B\hat{z}$, one can choose a Landau gauge, $\mathbf{A} = (-By, 0, 0)$. Since with this choice, the Hamiltonian is independent of the spatial variable, x. So $[\mathcal{H}, p_x] = 0$, and hence p_x continues to be a good quantum number.

Further, the Hamiltonian in Eq. (4.3.2.2) is valley decoupled, that is, there are no matrix elements that connect the two valleys, namely $\mathbf{K}$ and $\mathbf{K}'$. Thus, it allows us to look at the solutions at each valley separately. For the $\mathbf{K}$ point, we have a coupled equation for the wavefunctions, ϕ^A and ϕ^B,

$$\epsilon \phi_A^K = v_F (p_x - ip_y)\phi_B^K \tag{4.3.2.3}$$

$$\epsilon \phi_B^K = v_F (p_x + ip_y)\phi_A^K. \tag{4.3.2.4}$$

One can insert Eq. (4.3.2.3) in Eq. (4.3.2.4) and Eq. (4.3.2.4) in Eq. (4.3.2.3) to obtain

$$\epsilon^2 \phi_A^K = \hbar^2 v_F^2 (p_x - ip_y)(p_x + ip_y)\phi_A^K \tag{4.3.2.5}$$

$$\epsilon^2 \phi_B^K = \hbar^2 v_F^2 (p_x + ip_y)(p_x - ip_y)\phi_B^K. \tag{4.3.2.6}$$

Inserting the Landau gauge such that $p_x \to p_x + eBy$,

$$\frac{\epsilon^2}{\hbar^2 v_F^2}\phi_B^K = (p_x + eBy + ip_y)(p_x + eBy - ip_y)\phi_B^K \tag{4.3.2.7}$$

$$= \left[(p_x + eBy)^2 - i[(p_x + eBy), p_y] + p_y^2\right]\phi_B^K.$$

Since $[p_x, p_y] = 0$ and $[y, p_y] = i\hbar$, one gets

$$\frac{\epsilon^2}{\hbar^2 v_F^2}\phi_B^K = [(p_x + eBy)^2 + e\hbar B + p_y^2]\phi_B^K. \tag{4.3.2.8}$$

Thus, we arrive at

$$\left(\frac{\epsilon^2}{\hbar^2 v_F^2} - e\hbar B\right)\phi_B^K = (\tilde{p}_x^2 + \tilde{p}_y^2)\phi_B^K, \tag{4.3.2.9}$$

where $\tilde{p}_x^2 = (p_x + eBy)^2$ and $\tilde{p}_y^2 = p_y^2$. Dividing both sides by $2m$,

$$\frac{1}{2m}\left(\frac{\epsilon^2}{\hbar^2 v_F^2} - e\hbar B\right)\phi_B^K = \left(\frac{\tilde{p}_x^2 + \tilde{p}_y^2}{2m}\right)\phi_B^K = \left(\frac{1}{2}\tilde{k}(y - y_0)^2 + \frac{\tilde{p}_y^2}{2m}\right)\phi_B^K \tag{4.3.2.10}$$

with $\tilde{k} = \frac{e^2 B^2}{m}$, $y_0 = \frac{p_x}{eB}$. Thus, the RHS is identified as the Hamiltonian for a particle executing SHM in two dimensions about a coordinate point $(0, y_0)$. Thus,

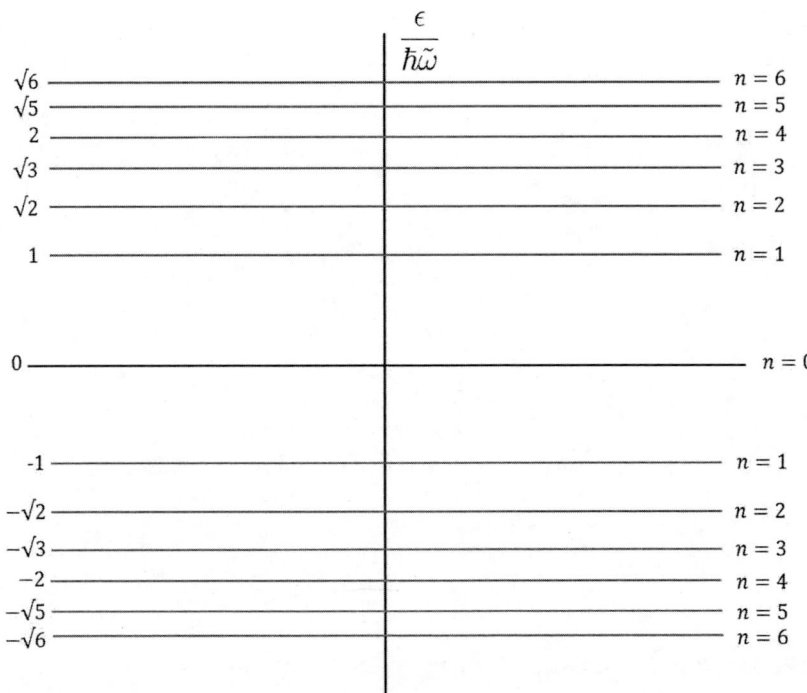

FIGURE 4.9. Plot showing the Landau levels in graphene for different indices, n.

it is obvious that the energy spectrum is given by $\epsilon_n = (n + \frac{1}{2})\hbar\omega_B$ with $\omega_B = \frac{eB}{m}$. Hence,

$$\frac{1}{m}\frac{\epsilon^2}{\hbar^2 v_F^2} = 2\left(n + \frac{1}{2}\right)\hbar\omega_B + \hbar\omega_B = \frac{2n\hbar\omega_B}{m}, \qquad \text{where } n = 0, 1, 2, \ldots \quad (4.3.2.11)$$

Eq. (4.3.2.11) allows positive and negative roots for ϵ. So we can obtain the energy spectrum as

$$\epsilon = sgn(n)\sqrt{n}\, v_F \frac{(2\hbar eB)^{1/2}}{m}. \qquad (4.3.2.12)$$

The energy spectrum is shown in Fig. 4.9. Let us define another quantity $\tilde{\omega} = v_F (2\hbar eB)^{1/2}$, so as to formally write the energy expression as that of a harmonic oscillator. We re-write the above expression as

$$\epsilon = \hbar\tilde{\omega}\, sgn(n)\sqrt{|n|}. \qquad (4.3.2.13)$$

Thus, as opposed to the familiar harmonic oscillators when n can take positive integer values (including zero), however, here in graphene all integers, that is, both positive and negative numbers, are allowed. The positive integers denote particles (or electrons) in the conduction band and the negative ones denote holes in the valence band. Further, unlike the 2DEG, here the Landau levels are not equidistant. The largest separation occurs between the lowest Landau level ($n = 0$) and the first one ($n = \pm 1$). This large gap essentially facilitates observation of QHE in graphene at large temperatures, which is even true for room temperature.[3]

So far we have been discussing spinless particles. Including the spin, there will be an additional two-fold degeneracy of the Landau levels owing to Zeeman spitting. A hierarchy of energy scales needs to be ascertained here. Let us compare the energy gap between the two lowest Landau levels and the Zeeman splitting corresponding to a typical magnetic field, B, for example, $B = 10T$.

$$\triangle E_{\text{LL}} = \frac{\hbar \omega_B / 2}{v_F \sqrt{e \hbar B}} \qquad \text{(for the successive Landau levels)} \qquad (4.3.2.14)$$

$$\triangle E_z = \sqrt{e \hbar B} \qquad \text{(for the Zeeman term).} \qquad (4.3.2.15)$$

For typical $v_F \simeq c/300$ (c: speed of light),

$$\frac{\Delta E_z}{\Delta E_{\text{LL}}} \simeq 10^2. \qquad (4.3.2.16)$$

Thus, the Zeeman energy scale is much larger than the Landau level splitting, which makes it imperative to include spin degeneracy. Thus, including the Zeeman term, the energy can be written as

$$\frac{\epsilon^2}{v_F^2} = 2\hbar eB(n + 1) \qquad n = 0, 1, \ldots, \qquad (4.3.2.17)$$

where the additional term in the RHS (denoted by $2\hbar eB$) accounts for the spin. The spacing between the consecutive Landau levels is shown in Fig. 4.10. The energy level $\epsilon = 0$ is not present in the spectrum even for $n = 0$. This lowest Landau level is somewhat special in the following sense. The $n = 0$ level receives contribution from only one sublattice at each of the Dirac points. For example, 'A' sublattice contributes to the wave function at the Dirac point **K**, and 'B' sublattice contributes at **K'**. However, the $n \neq 0$ Landau levels have non-zero amplitudes at both the A and B sublattices.

---

[3]The title of the paper by Geim and Novoselov is 'Room temperature quantum Hall effect in Graphene'. See Ref. [71].

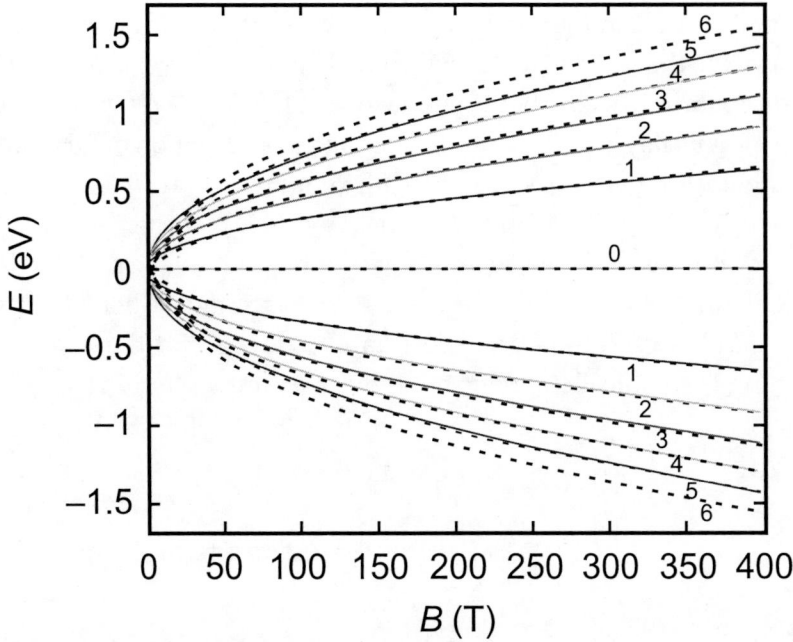

FIGURE 4.10. Plot showing the Landau levels for different indices, n, as a function of the magnetic field, B. The Landau levels vary as $\sqrt{n}$.

Finally, the wavefunctions corresponding to an arbitrary Landau level index at the two Dirac points, corresponding to the gauge we have chosen are

$$\psi_{n,k}^{K} = \frac{c_n}{\sqrt{L}} e^{-ikx} \begin{pmatrix} 0 \\ 0 \\ sgn(n)(-i)\phi_{|n|-1,k} \\ \phi_{|n|,k} \end{pmatrix} \tag{4.3.2.18}$$

and

$$\psi_{n,k}^{K'} = \frac{c_n}{\sqrt{L}} e^{-ikx} \begin{pmatrix} \phi_{|n|,k} \\ sgn(n)(-i)\phi_{|n|-1,k} \\ 0 \\ 0 \end{pmatrix} \tag{4.3.2.19}$$

with

$$\begin{aligned} c_n(x) &= 1 \quad \text{for } n = 0 \\ &= \frac{1}{\sqrt{2}} \quad \text{for } n \neq 0. \end{aligned} \tag{4.3.2.20}$$

Further,

$$sgn(n) = 0 \quad \text{for } n = 0$$
$$= \frac{n}{|n|} \quad \text{for } n \neq 0 \tag{4.3.2.21}$$

with

$$\phi_{n,k} = exp\left[-\frac{1}{2}\frac{(y - kl_B^2)^2}{l_B^2}\right] H_n\left[\frac{(y - kl_B^2)}{l_B}\right] \tag{4.3.2.22}$$

l_B is the magnetic length ($= \sqrt{\frac{\hbar}{eB}}$) as defined before, and $H_n(x)$ are the Hermite polynomials. $\phi_{n,k}$ denotes the eigenfunctions of an electron in the presence of a magnetic field. n refers to the Landau level index. The Landau levels for different values of the flux, ϕ, are shown in Fig. 4.11 as a function of one of the wavevector,

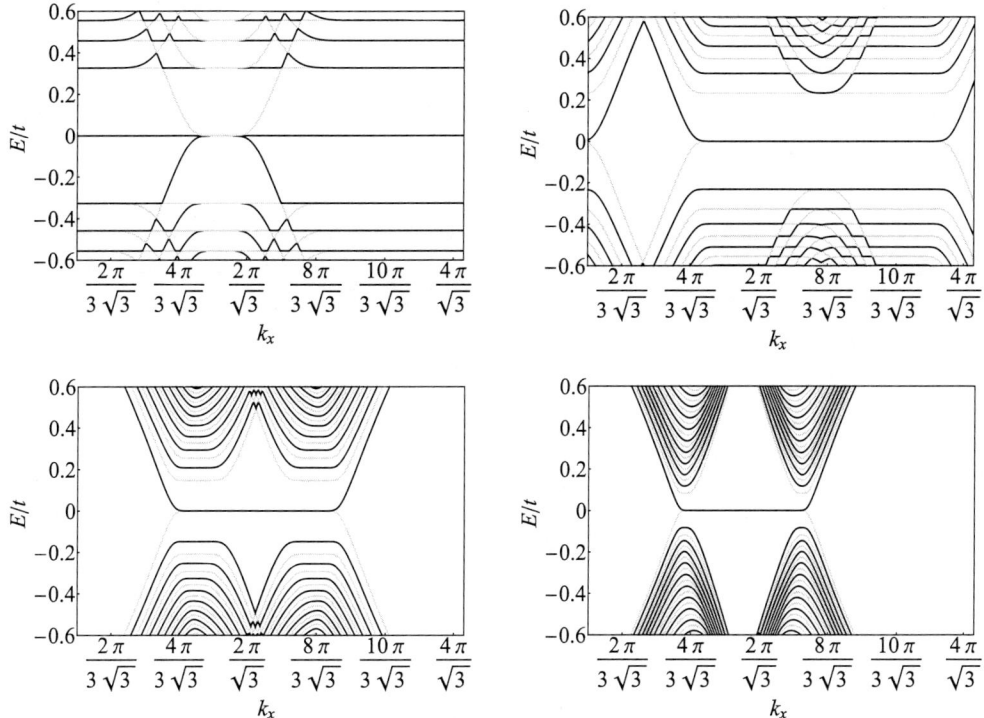

FIGURE 4.11. Plot showing the Landau levels in graphene for different values of flux, ϕ. The values of the fluxes are $\phi = \frac{\Phi_0}{100}$ (for the left upper panel), $\phi = \frac{\Phi_0}{200}$ (for the right upper panel), $\phi = \frac{\Phi_0}{500}$ (for the left lower panel) and $\phi = \frac{\Phi_0}{1600}$ (for the right lower panel). Here $\Phi_0 = \frac{h}{e}$ is the flux quantum.

k_x. As the magnitude of the flux is decreased, the width of the flat band appearing at the Fermi energy (at $E/t = 0$) decreases. Further, the flat bands become dissipative in the bulk corresponding to larger values of the Landau level index, n, and lower values of the flux ϕ. For a weak magnetic field, such that $\Phi/\Phi_0 = 1/1600$, the energy bands of the bulk regain their Dirac-like structure similar to the single particle energy levels, while the zero mode flat band continues to exist.

4.4   Hall conductivity of a graphene nanoribbon

The electronic properties of a graphene nanoribbon depend upon the geometry of the edges and the lateral width of the nanoribbon. According to the edge termination type of a nanoribbon, they are categorized as zigzag and armchair types (see Fig. 4.6). The electronic structure of the zigzag ribbon shows metallicity with zero band gap, while the armchair ones are conditionally metallic in the following sense. They are metallic when the lateral width W matches with $3M - 1$ (M being an integer), else they are semiconducting in nature with a finite band gap.

We have included numeric computation of the Hall conductivity for a zigzag nanoribbon of square shape with number of unit cells in each of the x and y directions to be 5120 as a function of the Fermi energy at certain values of the magnetic field, namely $B = 30T$ and $50T$ using Kubo formula (discussed in Chapter 1) in Fig. 4.12. The quantization of the Hall conductivity is clearly visible. Further, the longitudinal conductivity, σ_{xx}, is shown for one of them, namely $50T$. σ_{xx} shows vanishingly small values corresponding to plateaus of the Hall conductivities, while it shows spikes when the Hall conductivity jumps from one plateau to another. These features are consistent with the discussions made earlier. There is also an LL peak at zero energy which can be observed from the DOS (see Fig. 4.13) plotted at $B = 50T$. The presence of a zero-energy peak for this case is related to the chiral anomaly present.

It may be noted that the values of the magnetic field used here are quite high. In general, higher values of the magnetic field are required for relatively small system sizes for an efficient implementation of the numerical scheme of solving the Kubo formula.

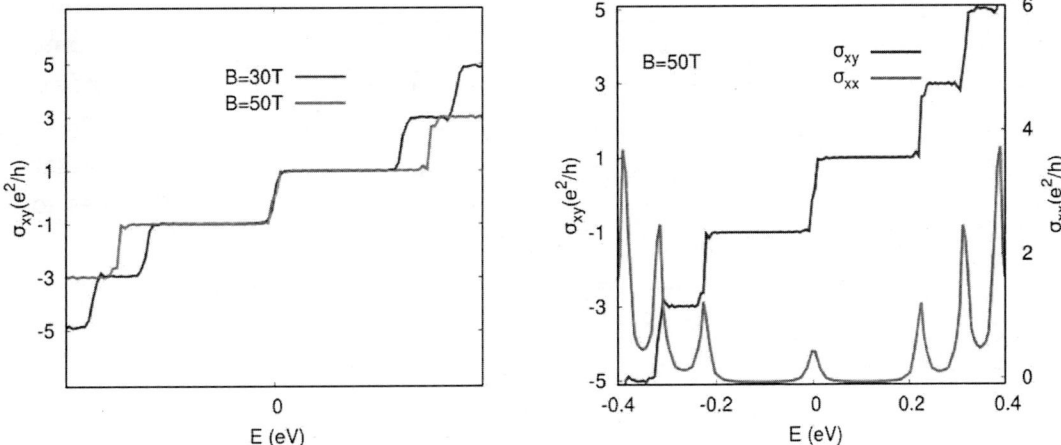

FIGURE 4.12. Plot showing the variation of Hall conductivity as a function of the bias voltage for a particular value of the magnetic field, such as $B = 30T$ and $50T$. The plateaus in the Hall conductivity are clearly visible. The longitudinal conductivity (σ_{xx}) is shown for a specific value, namely $B = 50T$. The ribbon consists of $L_x = L_y = 5120$ number of unit cells.

4.5 Experimental observation of the Landau levels in graphene

There are primarily two experimental techniques for observing the existence of Landau levels. They are (i) the infra-red (IR) spectroscopy [72] and (ii) scanning tunnelling microscopy (STM) [73] experiments. In the following discussion, we include a brief discussion on each of them and their utilities in observing the non-equidistant Landau levels in graphene.

In the IR spectroscopy, the optical transitions from one Landau level to another are studied via measuring the cyclotron frequencies. The Landau levels being proportional to $\sqrt{n}$ (n: Landau level index), all the frequencies of the optical transitions are distinct as the energy spacing between the consecutive Landau levels is not constant. These optical transitions are of two types which correspond to transitions between the electron or the hole states in the conduction or the valence bands (intra-band transition), or the transitions between the electron and the hole states pertaining to the valence and conduction bands (inter-band transition) respectively. The photoconductive response and the resistive voltage show the existence of differently spaced Landau levels in graphene at particular values of the magnetic field, longitudinal current and the IR frequency. The photoconductive intensity as a function of the carrier density, n (not to be confused

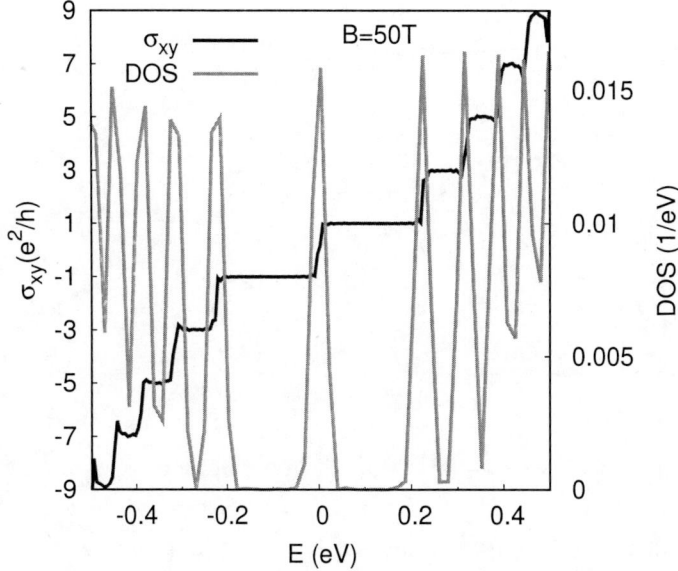

FIGURE 4.13. Plot for the Hall conductivity (the one with plateaus) and DOS (the one with spikes) are shown together for a zigzag ribbon. The zero energy peak is clearly visible from the DOS.

with the Landau level index) show distinct peaks which are proportional to the energy absorbed from the incident IR radiation.

In the STM experiment, the specific energy levels can be identified by varying the bias voltage between the tip and the surface of the sample, and the tunnelling current generated is proportional to the local density of states. In graphene, the Landau levels are directly observed via the peaks in the tunnelling spectrum. From the positions of the peaks as a function of the sample bias (shown in Fig. 4.14), the energies of the Landau levels can be extracted. Let us add a bit more detail on the experimental results presented in Fig. 4.14. The differential conductance, $\frac{dI}{dV}$, is plotted as a function of the bias voltage (in mV) in (a). Within a proportionality constant, it yields the local DOS, which is seen to vanish at zero bias, while it remains finite for graphite. The latter clearly implies the opening of a gap at the Dirac points as a single layer graphene is coupled to the substrate for the latter. Fig. 4.14(b) shows the surface map of the low-energy tunnelling conductance for two regions, one of which corresponds to the single layer graphene (darker region marked by 'A'), while the other (lighter region marked by 'B') denotes that of a graphite. In (c), the tunnelling spectra are shown as a function of the bias voltage for different values of the external field. The peaks correspond to the positions of the Landau levels. Finally, the massless nature of the Dirac fermions and the Landau

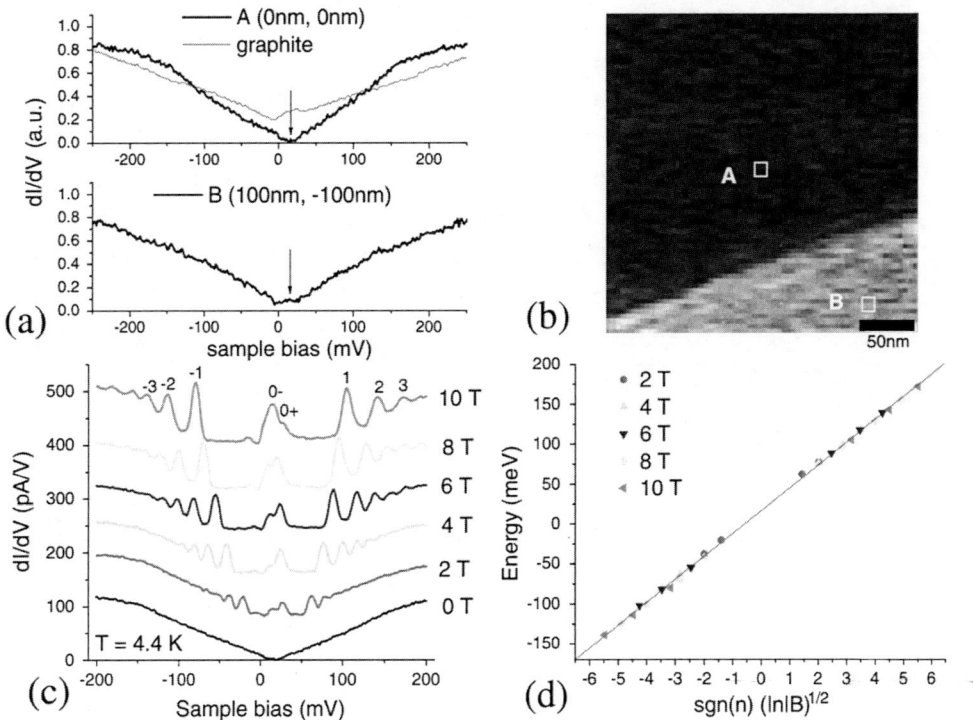

FIGURE 4.14. Plot of the scanning tunnelling spectra $\left(\frac{dI}{dV}\right)$ for graphene. (a) shows spectra at zero field taken in the regions 'A' and 'B' marked by squares in (b). A single layer graphene are shown by the black curve which vanishes at zero voltage. The spectra for graphite are shown for comparison. (b) shows the map of $\frac{dI}{dV}$ at energy as marked by arrows in (a). $\frac{dI}{dV}$ vanishes in the dark region, but is finite in the bright region. (c) shows the field dependence of the tunnelling spectra in the region 'A'. The peaks are labelled with LL index n. In (d), the energies of the Landau levels showing a square-root dependence on the level index, n (that is, $\sqrt{n}$) and for a few distinct field values. The symbols correspond to the peaks in (c). The figure is taken from Ref. [73].

level spacing varying as $\sqrt{n}$ are shown via plotting the energy, E, as a function of $\sqrt{|n|B}$ which allows the collapse of the data onto a single straight line shown in Fig. 4.14(d). The corresponding slope yields a fairly good estimate of the velocity of the Dirac fermions ($\sim 0.8 \times 10^6$ ms^{-1}). Such a collapse is obviously absent for graphite.

Now we shall wrap up the discussion of the QHE in monolayer graphene and give a brief introduction to the QHE in bilayer graphene in our subsequent discussion.

4.6 Bilayer graphene

It is established that the charge carriers in single layer graphene are massless chiral quasiparticles with linear (Dirac-like) dispersion which are eventually responsible for many unusual properties, including QHE at room temperature. On the other hand, a bilayer graphene is quite distinct than its single layer counterpart by the fact that it involves massive chiral quasiparticles with free electron-like (parabolic) dispersion [74] and hosts quadratic dispersion similar to 2DEG. Further, many of the properties, such as the electrical and the thermal conductivities, mobility, and so on of the bilayer system are higher than the single layer. The properties can also be efficiently tuned by doping or applying a gate voltage. Besides, owing to excellent mechanical strength, transparency to white light, and so on make them potential candidates for a large number of industrial applications. They include improved screen displays, thermoelectric devices, energy storage materials, and so on.

Another feature that is worth mentioning, although being beyond the scope of this book is the observation of superconductivity in a *twisted* bilayer graphene. One layer is rotated with respect to another by a certain angle, and at a particular *magic angle* the system promotes attractive pairing between the electrons, thereby enabling conduction with zero resistance. More significantly, the phase diagram at the magic angle as a function of carrier density resembles that of the high-T_c cuprates. A distinguishing feature of the phase diagram is the appearance of a non-Fermi liquid metal as the normal state characterized by linear in temperature (T) resistivity (instead of T^2). For details, the readers are encouraged to look at Ref. [75]. A recent publication [76] and the references therein are also useful.

We shall discuss QHE in bilayer graphene in the following section.

4.7 Quantum Hall effect in bilayer graphene

The quantum Hall effect in monolayer graphene has been discussed earlier and is a consequence of the chiral nature of the Dirac-like quasiparticles near certain points (the so-called Dirac points) in the BZ. In contrast, the low-energy Hamiltonian of bilayer graphene hosts a quadratic band structure (as if the linearly dispersing bands have split and have become quadratic near the Fermi surface because of the presence of the second layer). The chiral quasiparticles carry a Berry phase of 2π, instead of π for the single layer graphene. The spectral gap and the distinct electronic properties make the bilayer materials interesting from a research perspective.

We are of course interested in a strong perpendicular magnetic field in a bilayer graphene and contrast the scenario with regard to the Landau levels and quantized Hall plateaus with that of the monolayer. In the following discussion, we set our notations following Ref. [77] to write down the low-energy Hamiltonian of two coupled layers. There is a subtle point that deserves special mention. A bilayer stacking can in general be achieved by placing A sublattice points on top of A (and B on top of B), or in another arrangement A sublattice points are vertically stacked on top of B. The latter is called the Bernal stacking where a dimer bond (of strength, say, γ_1) is formed between B atom in layer 1 and A atom in layer 2 and vice versa. For obvious reasons, to incorporate the two layers, the size of the Hilbert space increases. One can use the basis $(c_{A_1}^\dagger, c_{B_2}^\dagger, c_{B_1}^\dagger, c_{A_2}^\dagger)$ for one of the Dirac points (say $\mathbf{K}$) and $(c_{B_2}^\dagger, c_{A_1}^\dagger, c_{B_1}^\dagger, c_{A_2}^\dagger)$ at the other Dirac point (namely $\mathbf{K'}$). Using $\lambda = \pm 1$ for $\mathbf{K}$ and $\mathbf{K'}$, the low-energy Hamiltonian for a bilayer, $\mathcal{H}_{BL}$, is written as

$$\mathcal{H}_{BL} = \begin{pmatrix} \lambda v_\perp \boldsymbol{\sigma} \cdot \mathbf{p} & \lambda v_F \boldsymbol{\sigma}^* \cdot \mathbf{p} \\ \lambda v_F \boldsymbol{\sigma} \cdot \mathbf{p} & \gamma_1 \sigma_x \end{pmatrix}, \tag{4.7.1}$$

where $v_F(= \frac{3at}{2})$ denotes the intra-layer nearest neighbour hopping, $\boldsymbol{\sigma} = (\sigma_x, \sigma_y, \sigma_z)$ are the Pauli matrices and refer to the sublattice degrees of freedom, $v_\perp$ is the velocity of the carriers across the layers ($v_\perp \ll v_F$) and γ_1 is a parameter which has a value $\gamma_1 \approx 0.39$ eV. The interlayer distance is about $c = 3.35$ Å.

In the presence of a magnetic field, the momentum transforms as

$$\mathbf{p} = -i\hbar\boldsymbol{\nabla} - e\mathbf{A},$$

where $\mathbf{A}$ denotes the vector potential corresponding to the transverse field, $\mathbf{B}$. New momentum operators are introduced, namely

$$\pi^+ = p_x - ip_y \; ; \quad \pi = p_x + ip_y \tag{4.7.2}$$

such that $[\pi, \pi^+] = 2\hbar eB$. Further, in a gauge, $\mathbf{A} = (0, Bx, 0)$, π^+ and π act on the eigenfunctions $|\phi_n\rangle$ as follows:

$$\pi^+|\phi_n\rangle = \frac{\sqrt{2}i\hbar}{l_B}\sqrt{n+1}|\phi_{n+1}\rangle, \quad \text{and} \quad \pi|\phi_n\rangle = \frac{\sqrt{2}i\hbar}{l_B}\sqrt{n}|\phi_{n-1}\rangle,$$

where $|\phi_n\rangle = e^{iky}|\phi_n(x)\rangle$ and $l_B = \sqrt{\frac{\hbar}{eB}}$. Thus, the low-energy Hamiltonian in the presence of a magnetic field becomes

$$\mathcal{H}_{BL} = \begin{pmatrix} 0 & \lambda v_\perp \pi & 0 & \lambda v_F \pi^+ \\ \lambda v_\perp \pi^+ & 0 & \lambda v_F \pi & 0 \\ 0 & \lambda v_F \pi^+ & 0 & \gamma_1 \\ \lambda v_F \pi & 0 & \gamma_1 & 0 \end{pmatrix}.$$

In a 2×2 form (using Pauli matrices), the Hamiltonian can be expressed as

$$\mathcal{H}_{BL} = \begin{pmatrix} \lambda v_3 (\sigma_x p_x - \sigma_y p_y) & \lambda v_F (\sigma_x p_x + \sigma_y p_y) \\ \lambda v_F (\sigma_x p_x + \sigma_y p_y) & \gamma_1 \sigma_x \end{pmatrix}. \tag{4.7.3}$$

The Landau level spectrum is given by [78,79]

$$E_n = \sqrt{n(n-1)} \hbar \omega_B, \quad n = 0, 1, 2 \ldots \tag{4.7.4}$$

The spectrum is still linear in the magnetic field, B, similar to the single layer case, but there is a subtle difference. Now the spectrum contains an $n = 0$ level which is independent of B. Moreover, the degeneracy of the $n = 0$ level is twice that of the $n \neq 0$ level.

There is an alternate form for the Landau levels used in the literature [80], namely

$$E_n = \text{sgn}(n) \sqrt{|n|(|n|+1)} \hbar \omega_B, \quad n = \ldots -2, -1, 0, 1, 2 \ldots \tag{4.7.5}$$

This form is more convenient as it yields the doubly degenerate zero energy Landau level. With this, the Hall conductivity is given by

$$\sigma_{xy} = \nu \frac{e^2}{h}, \quad \text{for } \nu = 4n, \quad n = \pm 1, \pm 2, \pm 3 \ldots \tag{4.7.6}$$

as opposed to $\nu = (4n+2)$ for a monolayer. The charge carriers are still chiral, but they behave like massive particles near the zero energy due to coupling between the layers. Since two orbital states possess zero energy, the zero energy mode is eight-fold degenerate. Whereas all the other Landau levels are four-fold degenerate which is similar to that of a monolayer.

There are further interests in the quantized Hall effect in bilayer systems when the two layers are slightly rotated with respect to each other (by a few degrees) compared to the perfect Bernal stacking. Such a torsion affects the electronic coupling between the layers and is bound to influence the Landau level spectrum. When one layer is twisted by a small angle θ, the distance between the Dirac points becomes a function of the rotation angle between the layers [81]. The linear dispersion is preserved and is confirmed via Raman studies [82] and is similar to the monolayer graphene. Also the eight-fold degeneracy which is the characteristic of a bilayer is topologically protected, but is sensitive to the rotation angle between the layers. The scenario is identical to the strained bilayer graphene as well [83].

5

Graphene as a Topological Insulator: Anomalous Hall Effect

5.1 Introduction

Having studied a prototype model Hamiltonian in one-dimensional (1D), we turn our focus towards two-dimensional (2D), now with the lens on graphene. Particularly, we shall explore whether graphene possesses the credibility of becoming a topological insulator. That may happen, provided by some means, we are able to open a spectral gap at the Dirac cones. Since a non-zero Berry phase can be a smoking gun for non-trivial properties, let us first look at the Berry phase of graphene.

5.1.1 Berry phase in graphene

For computing the Berry phase, let us use the low-energy Hamiltonian of graphene given by[1]

$$\mathcal{H} = \hbar v_F (\tau_z \sigma_x q_x + \sigma_y q_y), \tag{5.1.1.1}$$

where $\tau_z = +1$ and -1 represent the valleys K and K' respectively. As usual, σ_i are the 2×2 Pauli matrices that denote the sublattice degree of freedom.

---

[1]We have discussed the electronic properties of graphene in Chapter 4.

To remind ourselves, the integration of the Berry curvature over the first Brillouin zone (BZ) yields Chern number.

$$n = \frac{1}{2\pi} \oint_{BZ} \mathcal{F} d^2 K = C \quad \text{(a notation we have used earlier)}.$$

We rewrite the Dirac Hamiltonian as

$$h(\mathbf{q}) = \mathbf{q} \cdot \boldsymbol{\sigma}. \tag{5.1.1.2}$$

In the above equation, we have dropped the velocity term for simplicity. In the polar coordinate $\mathbf{q}$ and $h(\mathbf{q})$ are written as follows

$$\mathbf{q} = |\mathbf{q}| \begin{pmatrix} \cos\phi \\ \sin\phi \end{pmatrix} = q \begin{pmatrix} \cos\phi \\ \sin\phi \end{pmatrix} \tag{5.1.1.3}$$

and

$$h(\mathbf{q}) = q \begin{pmatrix} 0 & \cos\phi - i\sin\phi \\ \cos\phi + i\sin\phi & 0 \end{pmatrix} = q \begin{pmatrix} 0 & e^{-i\phi} \\ e^{i\phi} & 0 \end{pmatrix}. \tag{5.1.1.4}$$

The normalized eigenvectors are

$$|\psi_-\rangle = \frac{1}{\sqrt{2}} \begin{pmatrix} -e^{-i\phi} \\ 1 \end{pmatrix} \quad \text{and}$$

$$|\psi_+\rangle = \frac{1}{\sqrt{2}} \begin{pmatrix} e^{-i\phi} \\ 1 \end{pmatrix}. \tag{5.1.1.5}$$

Next, we calculate the Berry connection $\mathcal{A}$, and to remind ourselves only that filled bands are to be taken into account. So we shall consider $|\psi_-\rangle$ in the definition of $\mathcal{A}$.

$$\mathcal{A} = i \langle \psi_- | \nabla_q | \psi_- \rangle. \tag{5.1.1.6}$$

Here ∇_q is the gradient operator in polar coordinate which is given by

$$\nabla_q = \left(\frac{\partial}{\partial q} \hat{q} + \frac{1}{q} \frac{\partial}{\partial \phi} \hat{\phi} \right). \tag{5.1.1.7}$$

Note that $|\psi_-\rangle$ does not depend upon q. It should be noted that we can introduce a band index n corresponding to the Chern number of a band, that is, C_n. Therefore, the total Chern number can be written as the sum of the Chern numbers of all bands, which are,

$$C = \sum_n C_n$$

$$C_n = \frac{1}{2\pi} \int_S \mathcal{F}_n dS. \qquad (5.1.1.8)$$

Where S is the surface which encloses the loop. With $\mathcal{A} = \frac{1}{2q}$, $\nabla \times \mathcal{A} = 0$. So $\mathcal{F} = 0$, and hence $C = 0$ which is not a surprise, as for time reversal invariant systems the Chern number should vanish.

The winding number around the Dirac points signifies the Berry phase that has values $+\pi$ and $-\pi$ for the **K** and **K'** points. Therefore, it helps us to assign a topological *charge* of the Dirac points, which will further tell us about the winding of the wave functions around such singular points in **k**-space differently with respect to each other. The **K** point carries the topological charge $+1$ (a vortex) and the **K'** points carry a topological charge -1 (an anti-vortex). With the Dirac Fermion sitting at **K** carries a Berry phase, $\Phi_B^K = \pi$, and the Dirac Fermion at **K'** has a Berry phase, $\Phi_B^{K'} = -\pi$. The overall Berry phase, Φ_B, is zero, that is, $\Phi_B = 0$.

5.1.2 Symmetries of graphene

It is fairly well known to the readers by now that graphene is represented by the nearest neighbour tight-binding model on a honeycomb lattice with a two sublattice basis, namely A and B sublattices. Carbon (C) atoms occupy both the sublattices. The situation is slightly different in boron nitride, which in spite of possessing the same crystal structure, the sublattice symmetry is broken by (non-equivalent) boron and nitrogen occupying the A and B sublattices. Thus, graphene is the prototype of a system possessing sublattice symmetry, which renders the Hamiltonian block off diagonal written in the sublattice basis. The low-energy physics of this model is denoted by the massless Dirac Hamiltonian that we have seen at length earlier. Here, for the sake of completeness, we recall that the low-energy form for the Hamiltonian of graphene at both the Dirac points, namely **K** and **K'**, is written as

$$\mathcal{H}_0(\mathbf{k}) = \hbar v_F (k_x \sigma_x \tau_z + k_y \sigma_y), \qquad (5.1.2.1)$$

where the pseudospins σ_x, σ_y are the Pauli matrices, which represent the sublattice degrees of freedom, while the z-component of the Pauli matrix, namely τ_z distinguishes the valleys at **K** and **K'**. Needless to say here is that there is no spin in the Hamiltonian which remains a valid description till spin–orbit coupling (SOC) coupling is included. Now, we consider the inversion (or the sublattice) symmetry that switches the two sublattices and also alters the momentum $\mathbf{k}$ to $-\mathbf{k}$ (since $\mathbf{p} = m\frac{d\mathbf{r}}{dt}$). It implies that the $\mathbf{K}$ and $\mathbf{K}'$ valleys are interchanged under inversion. The corresponding operator that does this operation is given by

$$\mathcal{P} = \sigma_x \tau_x.$$

Under this inversion operator, the Hamiltonian transforms as

$$\mathcal{P}\mathcal{H}(\mathbf{k})\mathcal{P}^{-1} = \hbar v_F \, \sigma_x \, \tau_x \, (k_x \sigma_x \tau_z + k_y \sigma_y) \, \sigma_x \tau_x = \mathcal{H}_0(-\mathbf{k}). \tag{5.1.2.2}$$

The above relation can be proved by using product rules of the Pauli matrices, and it ensures inversion symmetry of the Dirac Hamiltonian.

Now we shall discuss time reversal symmetry. In the case of graphene, time reversal symmetry implies changing the momentum vector $\mathbf{k}$ to $-\mathbf{k}$, followed by complex conjugation of the operator (as explained earlier). Under the time reversal symmetry operation, one Dirac point (say, $\mathbf{K}$) goes to another one (say, $\mathbf{K}'$), and thus the two Dirac cones are exchanged. Thus, taking the time reversal operator, $\mathcal{T}$, to be a complex conjugation operator should have been sufficient. However, as discussed earlier, the time reversal symmetry in graphene also implies a transformation from one valley to another, that is, $\mathbf{K}$ changing over to $\mathbf{K}'$. This makes us settle for

$$\mathcal{T} = \tau_x \mathcal{K}, \tag{5.1.2.3}$$

where τ_x is the x-component of the Pauli matrix. Note that here $\mathcal{T}^2 = 1$ as we are dealing with spin-less fermions.[2] To check for the invariance of the Hamiltonian under the operation of $\mathcal{T}$, one needs to prove

$$\mathcal{T}\mathcal{H}_0(\mathbf{k})\mathcal{T}^{-1} = \hbar v_F \, \tau_x \, (k_x \sigma_x \tau_z + k_y \sigma_y^*)\tau_x = \mathcal{H}_0(-\mathbf{k}). \tag{5.1.2.4}$$

The above relation ensures the invariance of the Dirac Hamiltonian under the time reversal operation.

It may also be mentioned that the above symmetries put together, that is, the product of the sublattice (or inversion), and the time reversal symmetries yield a further discrete symmetry, known as the charge-conjugate symmetry, usually denoted by $\mathcal{C}$. It can be checked that $\mathcal{H}_0$ is invariant under the combination of these two symmetries.

---

[2]For spin-full systems, $\mathcal{T}^2 = -1$.

To summarize, in the context of graphene, we have seen the emergence of three discrete symmetries, namely the sublattice symmetry (or, the inversion symmetry, denoted by $\mathcal{P}$), the time reversal symmetry (denoted by $\mathcal{T}$) and, finally, a combination of the two, that is, the charge-conjugation symmetry ($\mathcal{C}$). They indeed have different properties, such as $\mathcal{P}$ is a unitary operator, and anticommutes with the Hamiltonian, $\mathcal{T}$ is an anti-unitary operator which commutes with the Hamiltonian, while the charge-conjugation operator $\mathcal{C}$ is anti-unitary (since it is a combination of $\mathcal{P}$ and $\mathcal{T}$), which also anticommutes with the Hamiltonian.

Having discussed the fundamental symmetries of graphene, let us return to its prospects of being a topological insulator. In Haldane's own submission (see https://topocondmat.org/w4_haldane/haldane_model.html), there can be simple efforts to tweak the Hamiltonian to achieve topological properties. Thus, the goal is to transform a sheet of graphene into a quantum Hall-like state whose bulk is an insulator but the edges are conductors. Further, the sheet must possess chiral edge modes, that is, the edge currents propagate in the opposite directions along the two edges of the ribbon. One could achieve such a scenario in two possible ways: either by breaking the inversion symmetry, keeping the time reversal symmetry intact or by breaking the time reversal symmetry, without altering the inversion symmetry. In the following section, we show that while the first option does not yield a topological phase, the second one indeed does. Nevertheless, we shall discuss both, which are, respectively known as the Semenoff insulator (obtained via breaking the inversion symmetry) and a Haldane (or a Chern) insulator (obtained via breaking the time reversal symmetry).

5.2   Semenoff insulator

In order to break the inversion symmetry, consider a staggered onsite potential of the form

$$\mathcal{H}' = \varepsilon_A \sum_{\mathbf{r}_A} c_A^\dagger(\mathbf{r}_A) c_A(\mathbf{r}_A) + \varepsilon_B \sum_{\mathbf{r}_B} c_B^\dagger(\mathbf{r}_B) c_B(\mathbf{r}_B), \tag{5.2.1}$$

where ε_A and ε_B are onsite potentials at sites A and B respectively. For $\varepsilon_A \neq \varepsilon_B$, the inversion symmetry (or the sublattice symmetry) is broken as is the case for the hexagonal boron nitride (h-BN), where the sites occupied by the C atoms in graphene are occupied by boron (B) and nitrogen (N) at the A and B sublattice sites, thereby causing the onsite energies to be unequal. Thus, including a term that has an equal and opposite magnitudes at the two sublattices, the low-energy

Hamiltonian can be written as

$$\mathcal{H}(\mathbf{q}) = \mathcal{H}_0(\mathbf{q}) + m_I \sigma_z, \tag{5.2.2}$$

where the m_I term makes the massless Dirac particle massive. Here, $m_I = (\varepsilon_A - \varepsilon_B)/2$ can be called the Semenoff mass. $m_I = 0$ for $\varepsilon_A = \varepsilon_B$. Further σ_z anticommutes with $\mathcal{H}_0(\mathbf{k})$, that is,

$$\{\sigma_z, \mathcal{H}_0\} = 0.$$

The spectrum is given by

$$E(\mathbf{q}) = \pm\sqrt{\hbar^2 v_F^2 q^2 + m_I^2}. \tag{5.2.3}$$

In a compact notation, one may write it as

$$E_\mu(q) = \mu\sqrt{\hbar^2 v_F^2 q^2 + m_I^2}, \tag{5.2.4}$$

where $\mu = \pm 1$ and each sign refers to a valley index. The spectrum is plotted in Fig. 5.1. Spectral gaps of magnitude $2m_I$ open up at each of the Dirac points. This insulator is known as the Semenoff insulator, whose nature of the spectral gap is trivial because of the following reasons. The gap vanishes as $m_I \rightarrow 0$.

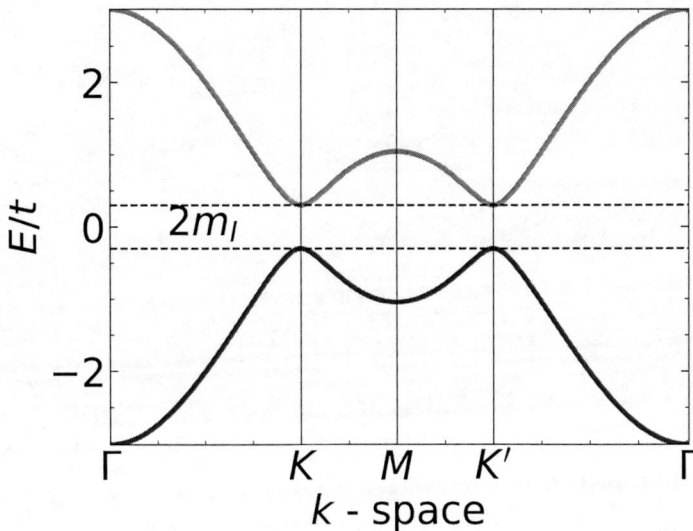

FIGURE 5.1. The electronic dispersion with a Semenoff mass, m_I, is shown. Gaps of magnitude $2m_I$ exist at the Dirac points.

Besides the band structure plotted for a graphene nanoribbon of size L_xL_y[3] shows no trace of the edge states. Further, the Chern number and the Berry phase also vanish as we shall show below, which certifies the trivial nature of the spectral gap. Hence, we cannot see any topological properties of the system in the presence of a non-zero m_I.

To gain a bit of detail on the Semenoff insulator, let us complete the mandatory calculations. The eigenfunctions can be written as

$$\Psi^\mu(\mathbf{q}) = \frac{1}{\sqrt{2}} \begin{pmatrix} \sqrt{1 + m_I/E^\mu} \\ \mu\sqrt{1 - m_I/E^\mu}\, e^{i\theta_q} \end{pmatrix}. \tag{5.2.5}$$

The corresponding Berry curvature is

$$\Omega^\mu = \frac{v_F^2 m_I}{2\mu \left[v_F^2 q_x^2 + v_F^2 q_y^2 + \beta^2 \right]^{\frac{3}{2}}}, \tag{5.2.6}$$

which eventually gives the Berry connection as (see Eq. 2.4.2)

$$\mathcal{A}^\mu = \frac{\tau_z}{2} \left(1 + \mu \frac{m_I}{|E^\mu|} \right) \frac{\hat{\theta}_q}{q}. \tag{5.2.7}$$

Finally, the Berry phase is obtained as

$$\Phi_B = \pi\tau_z \left(1 + \lambda \frac{m_I}{|E^\mu|} \right).$$

Thus, the Berry phase for a massless Dirac equation is renormalized by the Semenoff mass, m_I. One regains the corresponding result for graphene by putting $m_I = 0$.

A further (and more robust) check on the trivial nature of the spectral gap can be achieved by computing the dispersion for a graphene nanoribbon. A nanoribbon is a system that is infinite along one direction (say, x-direction), and finite along the other direction (y-direction). Usually, graphene ribbons are recognized by their edges along the x-axis, for example, with zigzag and armchair edges, and are referred to as the zigzag graphene nanoribbon (abbreviated as ZGNR) and armchair graphene nanoribbon (AGNR). There is an important difference between the two. ZGNR is always metallic with gapless edge states, while AGNR is

---

[3]This is called a semi-infinite ribbon. It is finite in the y-direction and very large (taken to be infinitely large) along the x-direction ($L_x \gg L_y$). We shall shortly discuss this later.

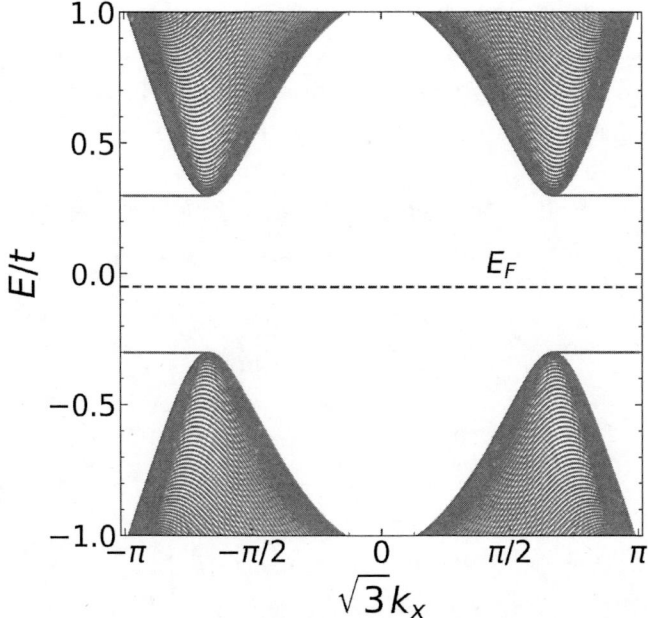

FIGURE 5.2.   The energy dispersion for a Semenoff insulator in a semi-infinite nanoribbon. A (trivial) gap is visible in the spectrum.

conditionally metallic in the following sense. AGNR has conducting edge states when $N = 3M - 1$, where N is the number of lattice sites in the y-direction, and M is an integer.

We have taken a ZGNR as shown in Fig. 5.3, with the total number of lattice sites along the y-axis as 256, that is, $N = 256$ (so the number of unit cells is 128) along the y-direction and a width given by $(\frac{3N}{2} - 1)a$ (a: lattice constant = 1.42Å), which upon putting $N = 256$ yields $383a$ or 543.86Å. We finally write down the equation of motion, that is, solving the Schrödinger equation, $\mathcal{H}\psi = E\psi$, for the amplitudes at the A and B sublattice sites as below,

$$E_k a_{k,n} = -\left[t\left\{ 1 + e^{(-1)^n ik} \right\} b_{k,n} + t b_{k,n-1} \right] + m_I a_{k,n} \tag{5.2.8}$$

$$E_k b_{k,n} = -\left[t\left\{ 1 + e^{(-1)^{n+1} ik} \right\} a_{k,n} + t a_{k,n+1} \right] - m_I b_{k,n}. \tag{5.2.9}$$

Along the x-direction, the ribbon is infinite, which is implemented in our numeric computation by assuming the momentum along the x-direction, that is, k_x to be a good quantum number. The above equations are numerically solved. We present the results in Fig. 5.2 which clearly show the absence of zero modes, which precludes its prospects as a candidate for a topological insulator.

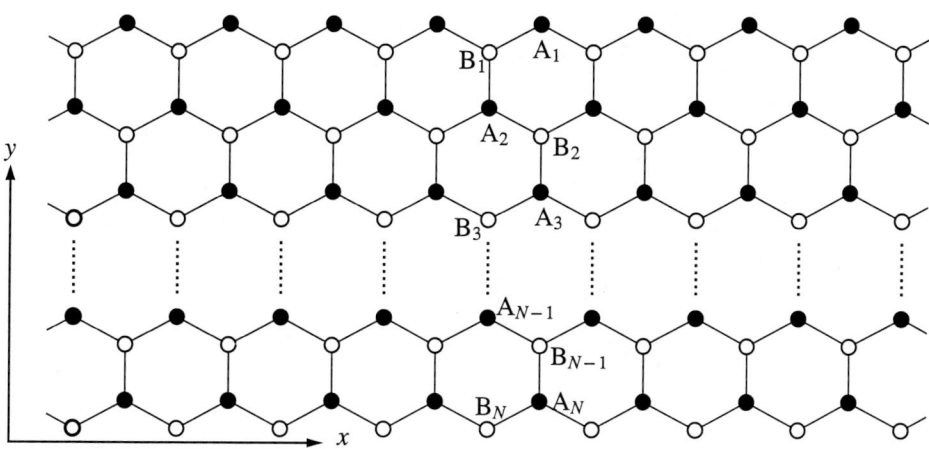

FIGURE 5.3. A schematic diagram of a semi-infinite zigzag nanoribbon is shown. We perform our numeric computation of the edge modes on a geometry such as this.

5.3 Haldane (Chern) insulator

The second option of breaking the time reversal invariance is more subtle, and yields a success in obtaining a topological state. The idea involves including an imaginary second neighbour hopping that assumes opposite signs depending on the direction of hopping. For example, if an anticlockwise hopping (shown by the solid arrow in Fig. 5.4) is assumed with a negative sign, then the clockwise hopping (shown by the dashed line in Fig. 5.4) appears with a positive sign. A formal way of writing this term is via

$$\mathcal{H}'' = t_2 \sum_{\langle\langle ij \rangle\rangle} e^{i \nu_{ij} \phi} c_i^\dagger c_j, \tag{5.3.1}$$

where the sum runs over the next nearest neighbour (NNN) sites (double angular bracket $\langle\langle ij \rangle\rangle$ imply NNN sites). ν_{ij} denotes the chiral nature of the hopping term where $\nu_{ij} = -\nu_{ji}$ depend on the direction of the hopping. The convention is $\nu_{ij} = +1$ for clockwise hopping between the NNN sites, while $\nu_{ij} = -1$ for anticlockwise hopping (see Fig. 5.4). Thus, the phase $e^{\pm i\phi}$ depends on the direction of the hopping that breaks the time reversal invariance because it flips the direction of the hopping. It should be noted that only the imaginary part of $e^{i\phi}$ is interesting. Hence, we take the value of ϕ to be $\pi/2$, which is known as the Haldane flux, and a presence of ϕ ascertains anomalous quantum Hall effect. It is anomalous in the sense that the Hall effect is realized without an external magnetic field, or equivalently, without the Landau levels. As we have seen earlier, and again shall

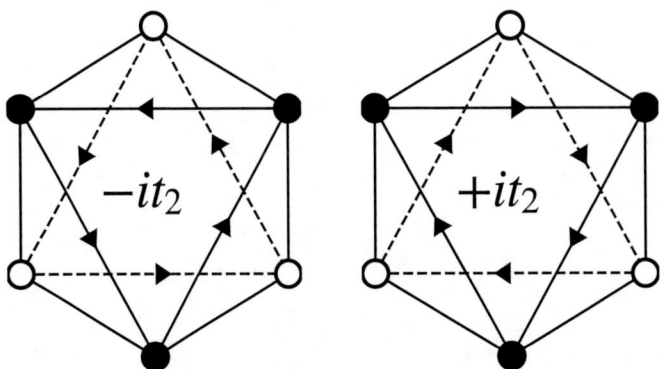

FIGURE 5.4.   The direction dependent complex phase of the next nearest neighbour hopping in the Haldane model is shown.

see shortly that broken time reversal symmetry implies a finite Chern number, and the system is called the Chern insulator. The complex phases can be realized by applying staggered magnetic fields pointing in opposite directions at the centre of the honeycomb lattice relative to that at the vertices.

Introducing the NNN vectors, $\mathbf{b}_i$, as earlier, where $\mathbf{b}_1 = \delta_2 - \delta_3, \mathbf{b}_2 = \delta_3 - \delta_1$ and $\mathbf{b}_3 = \delta_1 - \delta_2$ with δ_i denoting the vectors connecting NN sites, the Hamiltonian can be written as (the NN tight-binding term is also there, but not written here)

$$\mathcal{H}'' = t_2 \sum_{i=1}^{3} \left[e^{i\phi} \sum_{\mathbf{r}_A} c_A^\dagger(\mathbf{r}_A) c_A(\mathbf{r}_A + \mathbf{b}_i) + e^{-i\phi} \sum_{\mathbf{r}_B} c_B^\dagger(\mathbf{r}_B) c_B(\mathbf{r}_B + \mathbf{b}_i) \right]. \quad (5.3.2)$$

The full tight-binding Hamiltonian in the momentum space reads

$$\mathcal{H}''(\mathbf{k}) = 2t_2 \left[\cos\phi \sum_{i=1}^{3} \cos(\mathbf{k} \cdot \mathbf{b}_i) \mathbb{1} + \sin\phi \sum_{i=1}^{3} \sin(\mathbf{k} \cdot \mathbf{b}_i) \sigma_z \right]. \quad (5.3.3)$$

Since the NNN Hamiltonian is a function of the wave vector $\mathbf{k}$, it is dispersive in nature. However, the low-energy Hamiltonian about the $\mathbf{K}$ and $\mathbf{K}'$ points is independent of $\mathbf{k}$ at the leading order, where the Hamiltonian can be shown to assume the form

$$\mathcal{H}''_{\pm \mathbf{K}} = m_H \, \tau_z \, \sigma_z, \quad (5.3.4)$$

where $m_H = -3\sqrt{3}\, t_2 \, \sin\phi$, and τ_z has values $+1$ and -1 which represent the $\mathbf{K}$ and the $\mathbf{K}'$ valleys respectively. The expression for m_H can be obtained by substituting $\mathbf{k} = \mathbf{K}$ in the Hamiltonian in Eq. (5.3.3), which gives

$$\sum_{i=1}^{3} \cos(\mathbf{k} \cdot \mathbf{b}_i) = -\frac{3}{2} \quad \text{and} \quad \sum_{i=1}^{3} \sin(\mathbf{k} \cdot \mathbf{b}_i) = \mp\frac{3\sqrt{3}}{2}.$$

The readers are encouraged to fill up a few steps of algebra.

It should be noted that the term in Eq. (5.3.4) breaks the time reversal symmetry. The reason is the following. τ_z alters its sign under the time reversal operation since it denotes the valley degree of freedom, while σ_z does not change its sign. Thus, the energy spectrum opens up a gap at the Dirac points for specific values of the complex second neighbour hopping t_2. In fact, adding a small t_2 yields a situation similar to adding a small Semenoff mass, m_I. However, when t_2 exceeds a value of $\pm m_H/3\sqrt{3}$, the spectral gap at one of the two Dirac points (either $\mathbf{K}$ or $\mathbf{K}'$) closes, while it opens at the other Dirac point for one of the signs mentioned above, say $t_2 = m_H/3\sqrt{3}$. The reverse happens for $t_2 = -m_H/3\sqrt{3}$ where the gap closes at the former Dirac point, while opening at the other. We show this in Fig. 5.5.

In order to describe the topological properties, we repeat the same calculations as that of the Semenoff insulator. The eigenfunctions for the Haldane model can be written as

$$\Psi^{\mu}(\mathbf{q}) = \frac{1}{\sqrt{2}} \begin{pmatrix} \sqrt{1 + \beta/E^{\mu}} \\ \mu\sqrt{1 - \beta/E^{\mu}} e^{i\theta_q} \end{pmatrix}, \tag{5.3.5}$$

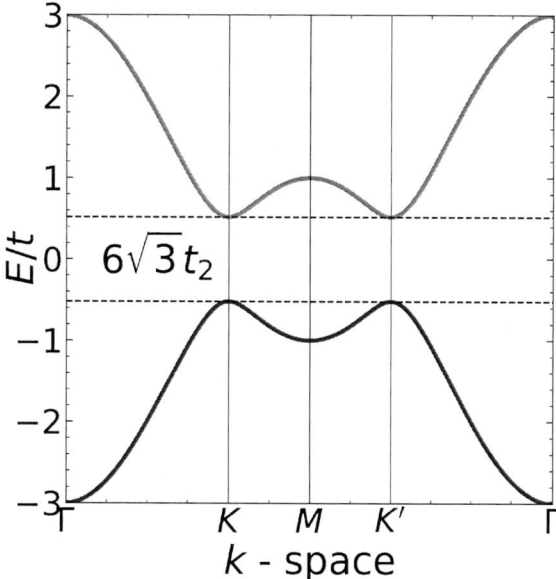

FIGURE 5.5. The energy spectrum of the Haldane model.

where $\mu = \pm 1$ and the energy spectrum yields

$$E^\mu = \mu\sqrt{v_F^2 q_x^2 + v_F^2 q_y^2 + \beta^2}.$$

This further yields a Berry curvature, which is given by

$$\Omega^\mu = \frac{v_F^2 \beta}{2\mu \left[v_F^2 q_x^2 + v_F^2 q_y^2 + \beta^2\right]^{\frac{3}{2}}}. \qquad (5.3.6)$$

The corresponding Berry connection has the following form

$$\mathcal{A}^\mu = \frac{\tau_z}{2}\left(1 + \mu\frac{\beta}{|E|^\mu}\right)\frac{\hat{\theta}_q}{q}, \qquad (5.3.7)$$

where $\beta = 3\sqrt{3}t_2$. The Berry phase, Φ_B, using

$$\Phi_B = \int \mathcal{A}^\mu.d\mathbf{q}$$

yields

$$\Phi_B = \pi\tau_z\left(1 + \lambda\frac{\beta}{|E^\mu|}\right). \qquad (5.3.8)$$

Finally, the Chern number can be obtained by integrating the Berry curvature over the BZ,

$$C = \oint \Omega^\mu(q)d^2q.$$

For the topological phase, that is, for $|t_2| > m_H/3\sqrt{3}$, one obtains a non-zero Chern number. The Chern number is the topological invariant which distinguishes the Semenoff insulator from a Chern insulator. For the Semenoff insulator, $C = 0$. We show the phase diagram in Fig. 5.6, where the topological phases are shown via the light shaded region ($C = 1$) and dark shaded region ($C = -1$) respectively, while the trivial region ($C = 0$) outside the lobes appears without shade. For completeness, in the appendix we describe a numerical method to compute the Chern number for a crystalline system. This goes by the name of Fukui's method [14].

How do we know that the nature of the gap is topological, instead of a trivial one as seen for a Semenoff insulator? This is a valid question since the nature of the gaps looks fairly the same in Figs. 5.2 and 5.5, except that the spectral gaps

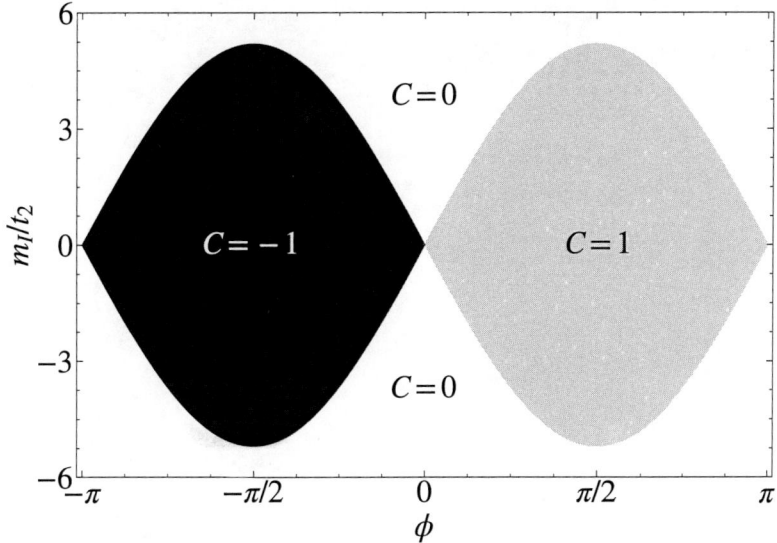

FIGURE 5.6. The Chern number phase diagram for the Haldane model. The light-shaded region corresponds to $C = 1$, while the dark-shaded region denotes $C = -1$. The region without any shade outside the lobes refers to a trivial insulator with $C = 0$.

carry the energy scales proportional to their *masses*, that is, m_I for the Semenoff insulator, and m_H for the Chern insulator. Further, to differentiate the nature of the spectral gap for the two cases, we discuss the band structure for a semi-infinite nanoribbon.

In a similar fashion as discussed in the context of a Semenoff insulator, the equations of the motion for the amplitudes at the A and B sublattice sites can now be written as

$$E_k a_{k,n} = -\left[t \left\{ 1 + e^{(-1)^n ik} \right\} b_{k,n} + t b_{k,n-1} \right]$$
$$- 2t_2 \left[\cos(k+\phi) a_{k,n} + e^{(-1)^n \frac{ik}{2}} \cos\left(\frac{k}{2} - \phi \right) \left\{ a_{k,n-1} + a_{k,n+1} \right\} \right] \quad (5.3.9)$$

$$E_k b_{k,n} = -\left[t \left\{ 1 + e^{(-1)^{n+1} ik} \right\} a_{k,n} + t a_{k,n+1} \right]$$
$$- 2t_2 \left[\cos(k-\phi) b_{k,n} + e^{(-1)^{n+1} \frac{ik}{2}} \cos\left(\frac{k}{2} + \phi \right) \left\{ a_{k,n-1} + a_{k,n+1} \right\} \right]. \quad (5.3.10)$$

In Fig. 5.7, we show the appearance of the edge modes as t_2 crosses $\pm m_H / 3\sqrt{3}$ which are absent at small values of t_2. The appearance of the edge modes implies the emergence of the topological phase in the model, and there occurs a phase transition from a topological insulating phase to that of a band insulator. Thus, we

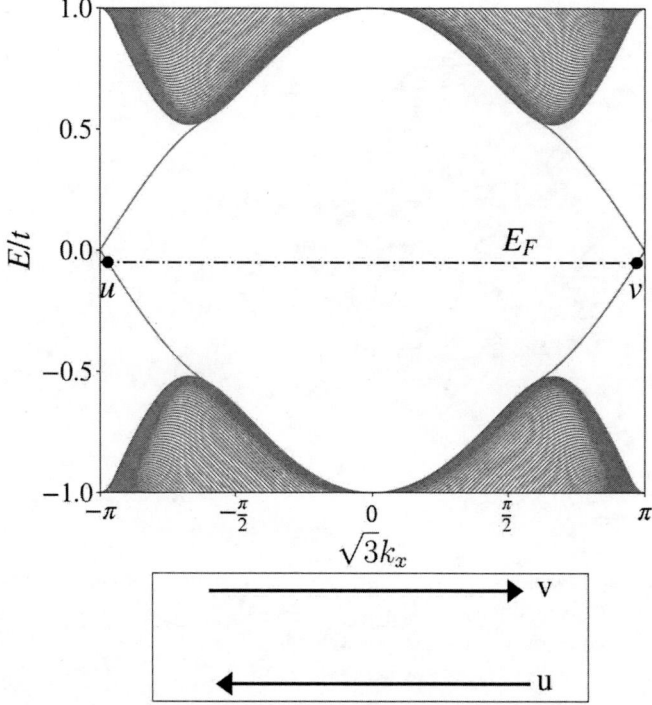

FIGURE 5.7.   The energy dispersion for a Chern insulator is a semi-infinite nanoribbon. The edge states are shown via solid lines that are split from the bulk bands. In the panel below, we show the chiral edge currents that flow in opposite directions at the points u and v along the edge modes.

get a quantum Hall-like state, with conducting edge modes (and insulating bulk), albeit without an external magnetic field.

5.4   Quantum anomalous Hall effect

Finally, we shall present the results for the Hall conductivity. Here, a non-zero Berry curvature yields a finite conductance. To remind ourselves, the full tight-binding Hamiltonian (including the NN term which we have excluded earlier in Eq. (5.3.3)) is written as [12]

$$
\mathcal{H} = -t \left[\cos(\mathbf{k} \cdot \boldsymbol{\delta}_1) + \sum_{i=2}^{3} \cos(\mathbf{k} \cdot \boldsymbol{\delta}_i) \right] \sigma_x - t \left[\sin(\mathbf{k} \cdot \boldsymbol{\delta}_1) + \sum_{i=2}^{3} \sin(\mathbf{k} \cdot \boldsymbol{\delta}_i) \right] \sigma_y
$$

$$
+ \left[\Delta - 2\, t_2\, \sin \phi \sum_{i=1}^{3} \sin(\mathbf{k} \cdot \boldsymbol{\nu}_i) \right] \sigma_z + \left[2t_2\, \cos \phi \sum_{i=1}^{3} \cos(\mathbf{k} \cdot \boldsymbol{\nu}_i) \right] I
$$

$$
= h_x \sigma_x + h_y \sigma_y + h_z \sigma_z + h_0 I, \tag{5.4.1}
$$

where h_x, h_y and h_z represent the coefficients of the Pauli matrices σ_i. The low-energy expansion of this Hamiltonian is convenient for our purpose. In fact, the computation of the Berry curvature is much easier for the low-energy Hamiltonian than it is for the full tight-binding one. Arriving at the low-energy Hamiltonian involves expanding the sine and the cosine functions to their leading order in the vicinity of the Dirac points. Applying these simplifications, one arrives at

$$\mathcal{H} = \mathbf{d} \cdot \boldsymbol{\sigma}, \tag{5.4.2}$$

where the components of the d-vector differ from those of the **h**-vector (see Eq. 5.4.1), and up to linear in the momenta k_x and k_y, they are given by

$$d_x(k_x, k_y) = \frac{3}{2}k_x, \quad d_y(k_x, k_y) = \frac{3}{2}k_y \quad \text{and} \quad d_z(k_x, k_y) = -3\sqrt{3}.$$

The above form facilitates computation of the Hall conductivity using the following form for the Berry connection [13],

$$\Omega(E_k) = \frac{\mathbf{d}}{2|\mathbf{d}|^3} \left(\frac{\partial \mathbf{d}}{\partial k_x} \times \frac{\partial \mathbf{d}}{\partial k_x} \right). \tag{5.4.3}$$

Finally, the Hall conductivity is obtained via

$$\sigma_{xy} = \frac{e^2}{h} \int \frac{d\mathbf{k}}{(2\pi)^2} f(E_k) \Omega(E_k), \tag{5.4.4}$$

where the integral is taken over the BZ, and $f(E_k)$ is the Fermi distribution function. Since our calculations are at zero temperature, so we set $f(E_k) = 1$. The Hall conductivity as a function of the Fermi energy is plotted in Fig. 5.8. A plateau at e^2/h is clearly visible which enunciates the quantization of the Hall conductivity. Also the Hall conductivity falls off symmetrically (owing to the particle–hole symmetry of the system) on either side of the zero Fermi energy. This happens due to the following reason. As the Fermi energy increases from zero, and it gets inside the conduction band, the Chern number of the conduction band starts gaining prominence. Being opposite in sign with respect to the Chern number of the valence band, the Chern numbers for both the bands start compensating for each other and eventually vanish, and so does the Hall conductivity when the Fermi energy reaches the top of the conduction band.

Further, the presence of only one plateau is confirmed by the value of the Chern number being 1 (or -1), and also that there is only a pair of gapless edge modes.

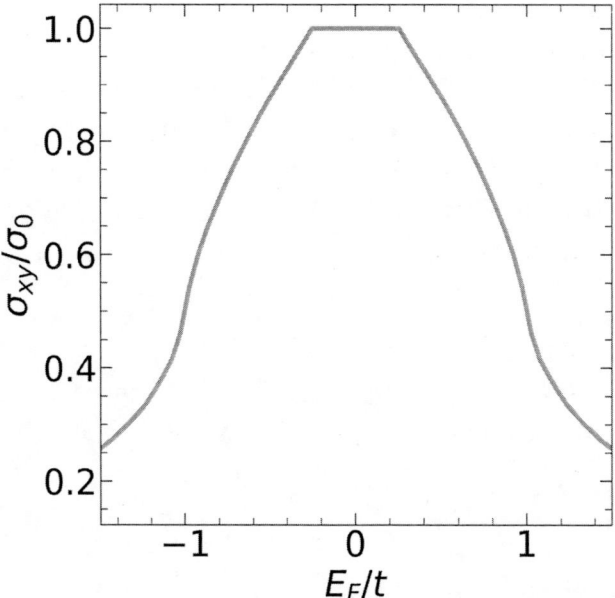

FIGURE 5.8.   The anomalous Hall conductivity is shown as a function of the Fermi energy. There is a distinct plateau in the vicinity of the zero Fermi energy.

Thus, an anomalous version of the Hall conductivity is indeed distinct from the usual Hall effect (in the presence of an external magnetic field). However, there are experimental realizations of systems with higher values of the Chern number, besides being backed up by a library of theoretical proposals.

5.5   Quantum spin Hall insulator

Let us set aside the complex second neighbour hopping due to Haldane for a moment; the Dirac points in graphene are protected by time reversal and the inversion symmetries. The complex second neighbour hopping among sites of the same sublattice breaks the time reversal symmetry as we have seen in the preceding section. C. L. Kane and E. G. Mele in 2005 [7] demonstrated that it is possible to restore the time reversal symmetry in the Haldane model if we include (real) spin in the Hamiltonian, thereby making two copies of the Haldane model, one for each spin. The inclusion of the spin opens up the possibility of an SOC, which, however, does not violate any of the fundamental symmetries that we have discussed above. Moreover, the resulting insulating phase is absolutely new and is referred to as the quantum spin Hall (QSH) phase. It should be clarified that SOC is not an essential ingredient for the realization of the QSH phase. However, to achieve a spin polarized transport in a material, which shall aid its usage for spintronic applications, SOC is essential. We shall return to this discussion shortly.

Similar to the quantum Hall phase, the QSH phase is distinct from the trivial insulators by the presence of the conducting states at the edges which are typically protected by the $\mathbb{Z}_2$ topological invariant. However, these edge states are non-chiral, unlike the quantum Hall states. In fact, they are called helical edge states, in the sense that there are two counter propagating edge modes at each edge, one for each spin (see Fig. 5.9). Such conducting modes are immune to single particle back scattering from defects, disorder or impurities as they are protected by the time reversal symmetry. Thus, as long as there is no time reversal symmetry breaking term, such as a magnetic impurity, and so on, the helical edge states are robust, and the QSH phase persists.

It was initially thought that graphene would host a QSH-like phase; however, it is almost impossible to realize such a phase owing to an extremely weak spin–orbit coupling [20]. However, a theoretical proposal of a QSH phase happened soon after when Bernevig, Hughes and Zhang [21] predicted that quantum wells made of CdTe/HgTe/CdTe host a QSH phase for a certain critical width of the HgTe layer where the band inversion occurs. The corresponding Hamiltonian is called the BHZ model (after Bernevig, Hughes and Zhang). Quite fortunately, immediately afterwards Molenkamp and co-workers [22] experimentally achieved such a scenario where an inverted band structure occurs, followed by the realization of the helical (instead of chiral) edge states.

In the following section, we shall describe the Kane–Mele model for graphene which serves as a toy model for a QSH phase that hosts counter propagating edge modes one for each spin at each of the edges. Further, these modes are found to be robust in the presence of a special kind of SOC, known as the Rashba spin–orbit coupling (RSOC). Here, as we shall show below, owing to the restoration of the time reversal symmetry, the Chern number is zero. However, the helical edge modes are still protected by bulk $\mathbb{Z}_2$ topological invariant, which is a consequence of Kramer's theorem applicable for band properties of fermions in a time reversal invariant system.

5.6 Kane–Mele model

The Dirac points have been shown to be protected by the time reversal and the inversion symmetries. If one of the symmetries get broken, the spectral gaps open at the Dirac points because of the splitting of degeneracy. Throughout our discussion on graphene thus far, the spin of the electron has not been incorporated in the calculation.[4] Kane and Mele have added the electron spin in the Hamiltonian of Haldane model which can be written as follows.

$$\mathcal{H} = -t \sum_{\langle ij \rangle, \alpha} c_{i\alpha}^{\dagger} c_{j\alpha} + it_2 \sum_{\langle\langle ij \rangle\rangle, \alpha\beta} v_{ij} c_{i\alpha}^{\dagger} (S_z)_{\alpha\beta} c_{j\beta} \tag{5.6.1}$$
$$= \mathcal{H}_0 + \mathcal{H}_{\text{KM}},$$

where $\mathcal{H}_0$ is the usual NN tight-binding term and $\mathcal{H}_{\text{KM}}$ is like the Haldane term, now summed over the (real) spins. It can be noticed from the second term that spin-↑ electron experiences a $\pi/2$ flux when it hops to the NNN site while the spin-↓ electron experiences a $-\pi/2$ flux for the same. Traditionally, this term is called the *intrinsic spin–orbit coupling*. S_z is the z-component of the spin of the electrons, and (α, β) denote the spin indices. S_z is indeed the z-component of the Pauli matrices $\boldsymbol{\sigma}$. However, to distinguish it from the sublattice and the valley indices, we have written it with S_z. Thus, the NNN hopping term (the second term) describes the coupling of the z-component of spin ($S_z = \pm 1$) with the chirality of the electrons, described by v_{ij}. It is as if the orbital angular momentum vector $\mathbf{L}$ is associated with the chirality in a familiar $\mathbf{L} \cdot \mathbf{S}$ term, thus justifying its identification as the SOC.

The second term, even though it resembles the Haldane term, respects all the symmetries of graphene. Time reversal flips the direction of hopping, that is, reversing the motion, but simultaneously it also flips the spin, thereby yielding another negative sign. This term respects all the symmetries of graphene. It may also be noted that the term does not involve spin flip, and hence the two bands of the Haldane model (one for each spin) behave distinctly. In a mathematical sense, it means that the Hamiltonian retains a block diagonal form, and hence is easy to deal with. The bands corresponding to the up-spin electrons are identical to the Haldane model (Chern insulators) discussed earlier. That is, they correspond to the phase of the complex NNN hopping to be $\phi = \pi/2$, and hence have opposite masses at the $\mathbf{K}$ and $\mathbf{K}'$ (remember the term $m_H \tau_z \sigma_z$ in the low-energy limit of the Haldane model). Further, it has a Chern number $C_\uparrow = +1$. Please note that we have brought in a spin index to the Chern number. For the down-spin electrons for which $\phi = -\pi/2$, there will be an extra negative sign, which implies reversed signs for m_H at the $\mathbf{K}$ and $\mathbf{K}'$ points as compared to the situation for the up-spin. This yields

---

[4]The Pauli matrices denote sublattice and valley degrees of freedom.

$C_\downarrow = -1$. Thus, the total Chern number, $\sum_\sigma C_\sigma = 0$, which is a consequence of the time reversal symmetry.

A simple way of seeing that the Kane–Mele model being two copies of the Haldane model is that

$$[\mathcal{H}_{KM}, S_z] = 0, \tag{5.6.2}$$

which signifies that $\mathcal{H}_{KM}$ decouples into Hamiltonian one for each spin. The situation is akin to a Haldane flux $\phi = \frac{\pi}{2}$ for one type of spin, and $\phi = -\frac{\pi}{2}$ for the other. The low-energy Kane–Male Hamiltonian can be shown to have a form (readers are encouraged to complete the derivation), written as

$$\mathcal{H}_{KM} = m_H \sigma_z \tau_z S_z,$$

where the amplitude $m_H = -3\sqrt{3}\, t_2$ is the Haldane mass as stated earlier.

Let us convince ourselves that time reversal is indeed a valid symmetry operation for the Kane–Mele model. For spinor particles, we have seen that the time reversal operator $\mathcal{T}$ is written as

$$\mathcal{T} = i\sigma_y K,$$

where K is a complex conjugation operator. Here, we write it as

$$\mathcal{T} = iS_y K.$$

However, since the time reversal transformations from one valley to another, an operator that can be implemented by incorporating a τ_x, we can write

$$\mathcal{T} = \tau_x\, iS_y K.$$

It is fairly trivial to see that $\mathcal{H}_{KM}$ is even under time reversal. Please recall that the time reversal inflicts complex conjugation, flips the real spin, reverses the valley degree of freedom, and in addition, reverses the direction of the momentum $\mathbf{k} \to -\mathbf{k}$. While the last one is not relevant, since the low-energy Hamiltonian is independent of $\mathbf{k}$, the first two yields under the time reversal,

$$\tau: \qquad \tau_z \to -\tau_z, \qquad s_z \to -s_z.$$

However, σ_z does not change sign as it denotes the sublattice degree of freedom. Hence, two negative signs cancel and we get $\mathcal{H}_{KM}$ to be even under $\mathcal{T}$.

To remind ourselves of the other fundamental symmetry, that is, the inversion symmetry $\mathcal{P}$, which yields

$$\mathcal{P}: \qquad \sigma_z \to -\sigma_z \qquad \tau_z \to -\tau_z, \qquad s_z \to s_z.$$

Hence, $\mathcal{H}_{KM}$ respects all symmetries of graphene as claimed earlier.

Let us look at the topological phase transition in a little more detail. For this, it is instructive to look at only one spin at a time, for example, $S_z = +1$ (that is, up spin). The Hamiltonian including a Semenoff mass becomes

$$\mathcal{H}_{KM}(\mathbf{k}) = \hbar v_F \left(k_x \sigma_x \tau_z + k_y \sigma_y \right) + (m_I + m_H \tau_z)\sigma_z. \tag{5.6.3}$$

Explicitly writing the above Hamiltonian for the two valleys,

$$\mathcal{H}_{KM}^{\mathbf{K}}(\mathbf{k}) = \hbar v_F (k_x \sigma_x + k_y \sigma_y) + (m_I + m_H)\sigma_z \tag{5.6.4}$$

$$\mathcal{H}_{KM}^{\mathbf{K'}}(\mathbf{k}) = \hbar v_F (-k_x \sigma_x + k_y \sigma_y) + (m_I - m_H)\sigma_z. \tag{5.6.5}$$

Now consider two possibilities: (i) $m_I > m_H$ and (ii) $m_H > m_I$. In the first case, consider the extreme limits (for convenience), that is, $m_I \gg m_H$ where we have a trivial band insulator. Now consider the other case where $m_H > m_I$: Nothing happens to $\mathcal{H}_{KM}^{\mathbf{K}}(\mathbf{k})$ in Eq. (5.6.5), but for $\mathcal{H}_{KM}^{\mathbf{K'}}(\mathbf{k})$ in Eq. (5.6.5), the gap closes and reopens. Thus, the insulating phase with $m_H > m_I$ is distinct from that of a band insulator by a *gap closing* phase transition, which by definition is a topological phase transition.

The situation for $S_z = -1$ is identical, except that the sign of m_H changes which results in a similar phase transition at the other Dirac point, that is, at the **K** point. Now defining,

$$\tilde{m} = m_I - m_H \tag{5.6.6}$$

yields, corresponding to,

$$\tilde{m} < 0, \qquad C = 1$$
$$\text{and} \qquad \tilde{m} > 0, \qquad C = 0. \tag{5.6.7}$$

5.7 Bulk–boundary correspondence

Bulk–boundary correspondence (BBC) yields a guide to the phenomenology of topological insulators. The topological invariants computed from the bulk properties corresponding to a particular phase of the system uniquely reflect the conducting edge modes. Let us try to answer the question that we have posed above, that is, how is $\tilde{m} < 0$ fundamentally different from that of $\tilde{m} > 0$? Again consider a semi-infinite nanoribbon, that is, infinite in the x-direction and finite in the y-direction. The Schrödinger equation with the Hamiltonian written earlier can now be solved for a semi-infinite system as we have already discussed.

Before we discuss the numerical solution for a nanoribbon, let us explore an analytic solution. We can assume that the Hamiltonian has an edge at $y = 0$, so that the system exists for $y < 0$ and a vacuum for $y > 0$. In addition, let us assume a particular value of k_x, namely $k_x = 0$ (remember k_x is a good quantum number owing to translational invariance in the x-direction). Hence, we can write down the Hamiltonian,

$$\mathcal{H}(y) = -iv_F\sigma_y\frac{\partial}{\partial y} + (m_I - m_H)\sigma_z, \tag{5.7.1}$$

where $\hbar = 1$ and $m_I - m_H = \tilde{m}(y)$. The RHS of Eq. (5.7.1) resembles a y-dependent potential energy in a 1D free Hamiltonian. Further, let us insist on

$$\begin{aligned} \tilde{m}(y) < 0, &\qquad \text{for } y < 0 \\ \tilde{m}(y) > 0, &\qquad \text{for } y > 0. \end{aligned} \tag{5.7.2}$$

Thus, there is a physical boundary between topological and trivial states. Let's look at the zero energy solution.

Now make an ansatz for the y-dependent wavefunction (like variational wavefunction)

$$\psi(y) = i\sigma_y \, e^{f(y)} \, \phi, \tag{5.7.3}$$

where ϕ is a 2-component spinor. Putting Eq. (5.7.3) in Eq. (5.7.1)

$$\left(iv_F\frac{df}{dy} + \tilde{m}(y)\sigma_x\right)\phi = 0 \qquad (\text{using} \quad \sigma_y\sigma_z = i\sigma_x). \tag{5.7.4}$$

The formal solution for $f(y)$ is obtained as

$$f(y) = -\frac{1}{v_F}\int_0^y dy' \, \tilde{m}(y'), \tag{5.7.5}$$

where ϕ is assumed to be eigenstate of σ_x with the eigenvalue $+1$.

Also the effect of $i\sigma_y = e^{i\frac{\pi}{2}\sigma_y}$ is to rotate by π around the y-axis.

$$\psi(y) = \exp\left(-\frac{1}{v_F}\int_0^y \tilde{m}(y')\,dy'\right)|\sigma_x = -1\rangle. \qquad (5.7.6)$$

The $\exp\left(-\frac{1}{v_F}\int_0^y \tilde{m}(y')\,dy'\right)$ factor allows it to fall off at the inside of the sample. So $\psi(y)$ the edge is maximum at the edges. Also, it is an eigenstate of σ_x as it has to mix the two sublattices by hopping along the boundary. For the other Dirac point, the state traverses in the other direction. At larger energies, $\epsilon(k_x) = -v_F k_x$, so that

$$v_F(\text{or } v) = \frac{\partial \epsilon(k_x)}{\partial x} = -v. \qquad (5.7.7)$$

For $\tilde{m} \to -\tilde{m}$, we have an electron traversing in the opposite direction at the other cone.

Finally, we show the numeric computation of the edge modes in Kane–Mele nanoribbon by solving the following sets of equations.

$$E_k a_{k,n} = \left[t\left\{1 + e^{(-1)^{n+1}ik}\right\}b_{k,n} + tb_{k,n+1}\right]s_0 + m_I a_{k,n}s_0$$

$$+ 2t_2\left[a_{k,n}\sin k + e^{(-1)^{n+1}\frac{ik}{2}}\sin\frac{k}{2}\{a_{k,n-1} + a_{k,n+1}\}\right]s_z$$

$$+ i\lambda_R\left[\left\{-\frac{1}{2}\left(1 + e^{(-1)^{n+1}ik}\right)b_{k,n} + b_{k,n+1}\right\}s_y\right.$$

$$\left. - \left\{(-1)^n\frac{\sqrt{3}}{2}\left(1 - e^{(-1)^{n+1}ik}\right)b_{k,n}\right\}s_x\right] \qquad (5.7.8)$$

$$E_k b_{k,n} = \left[t\left\{1 + e^{(-1)^n ik}\right\}a_{k,n} + ta_{k,n-1}\right]s_0 - m_I b_{k,n}s_0$$

$$+ 2t_2\left[b_{k,n}\cos k + e^{(-1)^n\frac{ik}{2}}\cos\frac{k}{2}\{a_{k,n-1} + a_{k,n+1}\}\right]s_z$$

$$+ i\lambda_R\left[\left\{\frac{1}{2}\left(1 + e^{(-1)^n ik}\right)b_{k,n} + b_{k,n-1}\right\}s_y\right.$$

$$\left. - \left\{(-1)^{n+1}\frac{\sqrt{3}}{2}\left(1 - e^{(-1)^n ik}\right)a_{k,n}\right\}s_x\right]. \qquad (5.7.9)$$

Fig. 5.9(a) clearly shows the existence of spin filtered edge modes in the topological phase, while they are absent in Fig. 5.9(b). The presence of the helical modes carrying spin resolved currents at each edge is shown in the panel below (in Fig. 5.9c). Each of those conducting modes denotes a channel for each spin. Thus,

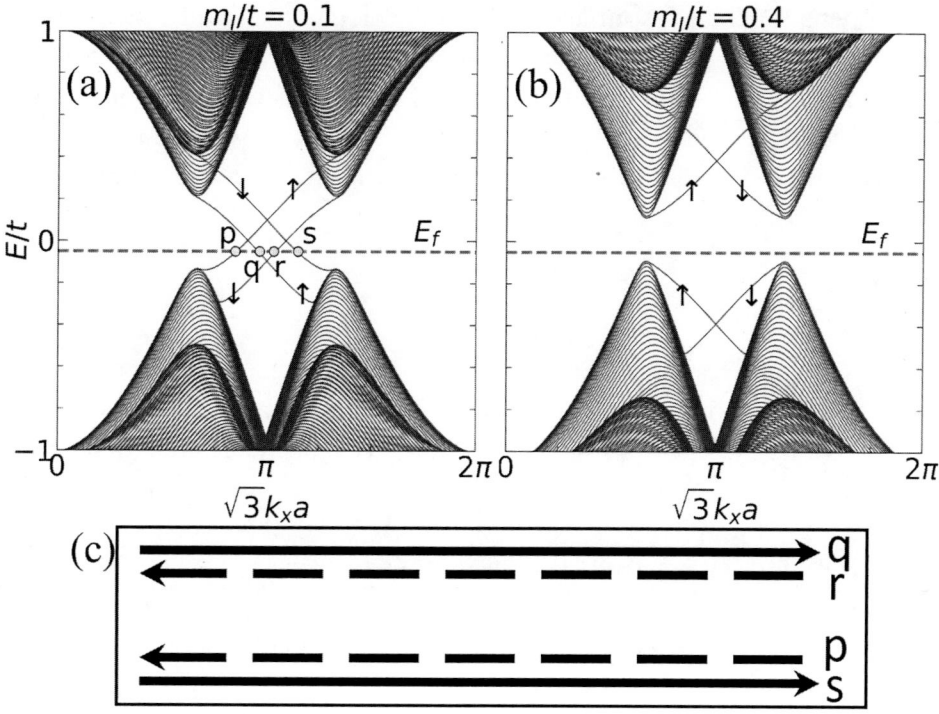

FIGURE 5.9. The helical edge modes for the Kane–Mele model on a semi-infinite nanoribbon showing (a) topological and (b) trivial phases. The panel below in (c) shows the spin polarized helical modes carrying current.

the model supports spin polarized conduction via the edge modes, while the bulk remains gapped. It must be kept in mind that the Chern number is identically equal to zero in this case owing to the time reversal symmetry being intact. However, the topological invariant here is the $\mathbb{Z}_2$ index which is non-zero.

5.8 Spin Hall conductivity

We have seen that even though the individual conducting edge states have a non-zero Chern number, the total Chern number, C, still vanishes owing to the time reversal symmetry being present. Thus, the charge Hall conductivity vanishes, that is, $\sigma_{xy} = 0$. However, the spin Hall conductivity survives.

In order to calculate the spin Hall conductivity, let us rewind the Corbino disc argument due to Laughlin. When a quantum of flux Φ_0 is added to the inner edge of the disc, an electron is transferred from an inner to an outer edge of the disc.

Say it happens for the up-spin leading to an e^2/h (charge) Hall conductivity. For the down-spin sector, in the presence of Φ_0, an electron is transferred backwards, that is from the outer edge to the inner one. Including both the spins, the total Hall conductivity is zero, as demanded by the time reversal invariance. However, in the process, a net spin is transferred from the inner to the outer edge. The corresponding spin Hall conductance is given by

$$G_s = \frac{\hbar}{2e}\left(\frac{e^2}{h} + \frac{e^2}{h}\right) = \frac{e}{2\pi}. \qquad (5.8.1)$$

Thus, G_s is quantized in unit of $e/2\pi$. The spin Hall conductance as a function of the Fermi energy is presented for graphene in Fig. 5.10. There a plateau in the vicinity of the zero Fermi energy (zero bias), a signature of the quantized nature of the spin Hall conductivity, is visible.

There is a subtle point about the intrinsic SOC that deserves special mention, namely the second term in Eq. (5.6.2) KM Hamiltonian commutes with the Hamiltonian, that is, $[\mathcal{H}_{KM}, S_z] = 0$. However, usually SOC terms are spin non-conserving which means they can mix different spins. Thus, in addition to the intrinsic SOC, other types of SOC can also be present. One such SOC is the RSOC, which, as we shall see later, mixes different spin components.

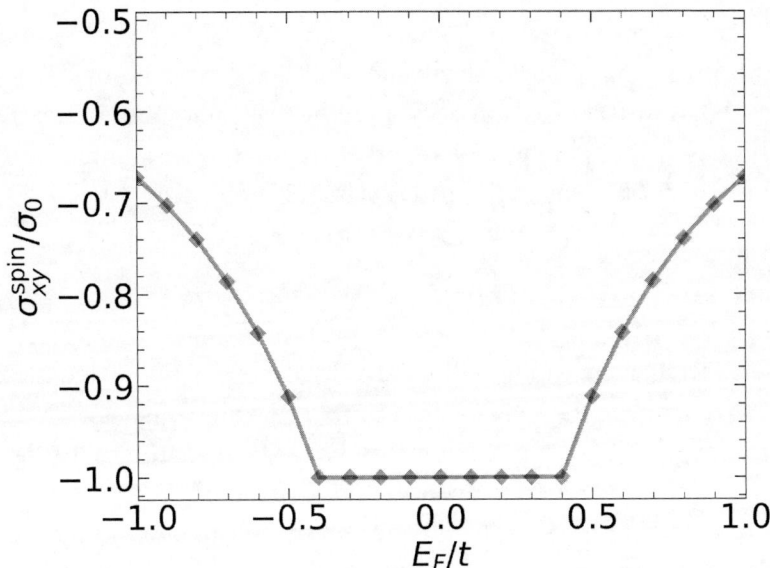

FIGURE 5.10. The spin Hall conductivity as a function of the Fermi energy is plotted. A quantization plateau in the vicinity of the zero Fermi energy is clearly visible.

5.9 Rashba spin–orbit coupling

In solids, free (or nearly free) electrons do not feel the strong attraction of the nucleus of their host atoms. However, the electrons still may experience an electric field or a potential gradient due to internal effects. As we know, if the electrons experience a strong electric field of potential gradient, then there is a possibility of emerging of an SOC. So if a potential gradient exists across the interface due to the structural inversion asymmetry, there will be an SOC, which is named after its discoverer, E.I. Rashba, that is, Rashba spin–orbit coupling, abbreviated as RSOC throughout our discussion [15]. The importance of the RSOC lies in the fact that asymmetry in the confinement potential can be varied by electrostatics means, allowing one to tune the RSOC strength by an external gate voltage. The strength of the RSOC depends also on the crystal structure in quantum wells, and is largest for narrow gap III − V semiconductors, such as InAs and InGaAs, and so on. In the following subsection, we shall describe the RSOC in a continuum model. Later on, we shall extend our discussion on graphene.

RSOC yields coupling of the wave vector of the electrons with their spin degrees of freedom. Further, it leads to the orientation of spins that points perpendicular to the direction of the electron propagation wave vector. The free particle Hamiltonian including RSOC is described by

$$\mathcal{H}_R = -\boldsymbol{\mu}.\mathbf{B} = -\boldsymbol{\mu}.\frac{\mathbf{v} \times \mathbf{E}}{c^2} \tag{5.9.1}$$

$$= \frac{eE}{mc^2}\mathbf{S} \cdot (\mathbf{v} \times \hat{\mathbf{z}}) = \frac{eE\hbar^2}{8\pi^2 m^2 c^2}\boldsymbol{\sigma}.(\mathbf{k} \times \hat{\mathbf{z}})$$

$$= \alpha_R(\hat{\mathbf{z}} \times \mathbf{k}) \cdot \boldsymbol{\sigma},$$

where $\alpha_R = \frac{eE\hbar^2}{8m^2\pi^2c^2}E$ is the strength of the RSOC, $\boldsymbol{\sigma}$ is a vector of Pauli spin matrices, $\mathbf{E} = -\nabla V$ is electric field along the $\hat{\mathbf{z}}$ direction. α_R can be tuned using an external gate voltage. In the absence of any Zeeman coupling, assuming elastic scattering and for $\hat{n} = \hat{z}$ (as per convention), the total Hamiltonian for the electron is given by

$$\mathcal{H} = \frac{p^2}{2m} + \alpha(\mathbf{p} \times \boldsymbol{\sigma}) \cdot \hat{z} = \frac{p^2}{2m} + \alpha(\sigma_x p_y - \sigma_y p_x). \tag{5.9.2}$$

This Hamiltonian yields the following energy spectrum,

$$E(k) = \frac{\hbar^2 k^2}{2m} \pm \alpha \, \hbar|k|, \tag{5.9.3}$$

where $|k|$ is the modulus of electron momentum with the plus and the minus signs denoting the two possible spin directions. The associated wave functions are given by

$$\Psi_{\pm}(x,y) = e^{i(k_x x + k_y y)} \frac{1}{\sqrt{2}} \begin{pmatrix} 1 \\ \pm i e^{-i\theta} \end{pmatrix}, \qquad (5.9.4)$$

where $\theta = tan^{-1}(k_y/k_x)$. It is easily understood that the spin states are always perpendicular to the direction of motion (Eq. 5.9.4). If an electron moves along the x-direction, the spinor part of the eigenvector becomes $(1, \pm i)$, that is, the spin up and the spin down are locked in the y-direction. By contrast, if the electron moves along the y-direction, the eigenvectors become $(1, \pm 1)$, that is, the spin up and the spin down states are constrained in the x direction (see Fig. 5.11). In Fig. 5.11(c–e), the energy spectrum as a function of momentum, k_y (keeping k_x constant) for a 2DEG are plotted corresponding to the following situations. Fig. 5.11(c) is related to a free electron in 2DEG where the spin degeneracy is present. Fig. 5.11(d) represents the energy spectrum for an electron in the presence of a magnetic field **B**, the spin degeneracy is lifted by the Zeeman splitting, and the gap separating spin up and spin down bands is equal to $g\mu_B B$ where g is the Bohr magneton. Fig. 5.11(e) presents a 1D view of the energy spectrum for an electron in the presence of RSOC.

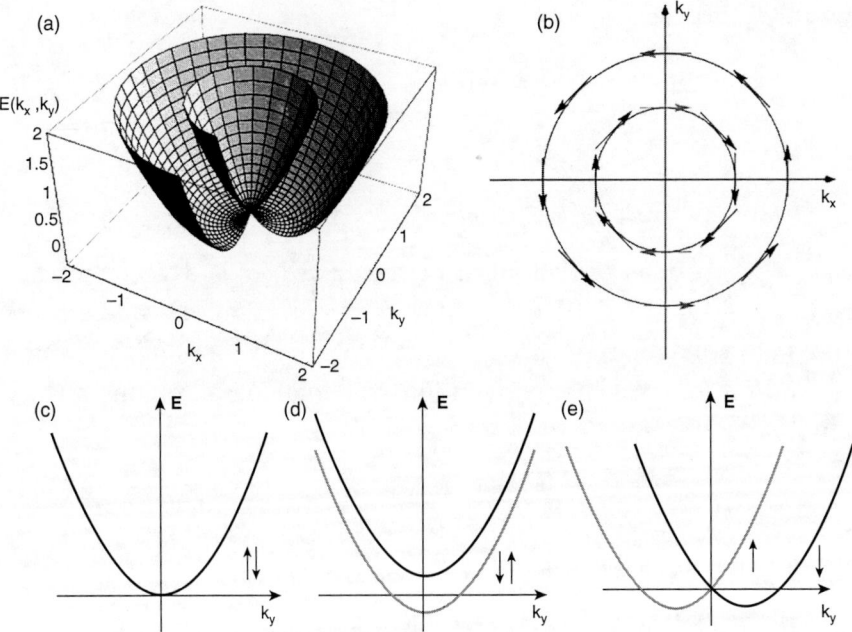

FIGURE 5.11.   (a) Three-dimensional energy spectrum of the Hamiltonian $\mathcal{H}$ (Eq. 5.9.2). (b) The Fermi energy contours for the Hamiltonian $\mathcal{H}$. (c) Energy spectrum for a free electron. (d) Energy spectrum for an electron in the presence of a magnetic field (Zeeman splitting). (e) Energy spectrum for an electron in the presence of Rashba spin–orbit coupling.

The spin degeneracy is lifted up except for $k_y = 0$. In this situation, the degeneracy is removed without opening any gap. At $k_y = 0$, the spin spectra are degenerate.

5.9.1 Rashba spin–orbit coupling in graphene

Writing the low-energy Hamiltonian in the vicinity of the Dirac points for graphene

$$\mathcal{H}_R = \lambda_R(S_x q_y - S_y q_x), \tag{5.9.1.1}$$

which mixes up and down spins, which is why $[\mathcal{H}_R, S_z] \neq 0$. We prefer to call the strength as λ_R here, instead of α_R which was earlier used by us. The corresponding spectrum is given by

$$E_{\gamma\delta}(\mathbf{q}) = \mu\sqrt{\mathbf{q}^2 + (m_H + \delta\lambda_R)^2} + \delta\lambda_R, \tag{5.9.1.2}$$

where the indices $\mu = \pm1$ and $\delta = \pm1$ yield the conduction and the valence band spectra at the $\mathbf{K}$ and $\mathbf{K}'$ points. As we have already realized that the spectrum is gapped in the presence of the intrinsic SOC (namely, the Haldane term) itself, and now when λ_R is included, it will start competing with m_H when $\delta = -1$. Also with increasing λ_R, the energy gap decreases. At $m_H = \lambda_R$, the spectrum consists of a Dirac cone with two gapped parabolic bands.

For the sake of completeness, we reiterate that the strength of RSOC is too weak to yield any observable effects. For example, $\lambda_R \sim 10^{-3}$ K in graphene, while the kinetic energy is much larger. There are techniques to enhance RSOC by using heavier adatoms, using an external gate voltage or bend the graphene layer. The basic idea is to create a strong gradient of the electric potential. We shall not discuss this any further, and suggest more specialized reviews on the subject.

5.10 Topological properties: The $\mathbb{Z}_2$ invariant

We have seen that the Kane–Mele model preserves the time reversal symmetry, and hence the Chern number, which is a $\mathbb{Z}$ invariant, would be zero. This requires us to look for a new topological invariant, namely the $\mathbb{Z}_2$ invariant which we shall discuss later. Now, on more general grounds we need to understand how the presence (or absence) of different discrete symmetries affect the topological invariant of a system. In the appendix, we include a discussion on the *tenfold* classification scheme that will aid us in decoding the nature of the topological invariant for a given system.

Since we shall include the Rashba SOC term in the KM Hamiltonian (see Eq. 5.6.1), and that it respects all symmetries of graphene (for example, Chern number equal to zero), a new topological invariant has to emerge. A priori, it is the $\mathbb{Z}_2$ invariant that we are talking about; however, we refer to the topological classification by Altland and Zirnbauer in Ref. [16, 17].

Let us discuss the topological invariant relevant here, namely the $\mathbb{Z}_2$ index that characterizes the topological properties of the system. We shall only talk about an inversion symmetric system. For the calculation of the $\mathbb{Z}_2$ index, one may consider the Bloch wave functions, $u_i(\mathbf{k}_i)$, of the occupied bands corresponding to a pair of points $\mathbf{k}_1$ and $\mathbf{k}_2$ in the BZ. These two points denote the locations of the band extrema (minima for the conduction band and maximum for the valence band) in the BZ. The wave function at one of these points can be obtained by time reversing the wave function corresponding to the other one, that is, $|u_i(\mathbf{k}_1)\rangle = \mathcal{T}|u_i(\mathbf{k}_2)\rangle$, and vice versa where $\mathcal{T}$ denotes the time reversal operator. Since the Hamiltonian is time reversal invariant, so we can decompose the Hamiltonian, $\mathcal{H}(\mathbf{k})$ and its corresponding occupied band wave functions, $|u_i(\mathbf{k})\rangle$, into even and odd subspaces. The even subspace has the property that $\mathcal{T}|u_i(\mathbf{k})\rangle$ is equivalent to $|u_i(\mathbf{k})\rangle$ up to a $U(2)$ rotation. Whereas the wave functions corresponding to the odd subspace have the property that the space spanned by $\mathcal{T}|u_i(\mathbf{k})\rangle$ is orthogonal to that of $|u_i(\mathbf{k})\rangle$. Now the $\mathbb{Z}_2$ invariant can be calculated by considering the momenta which belong to the odd subspace. We compute the expectation value of the time reversal operator between $|u_i(\mathbf{k})\rangle$ and $|u_j(\mathbf{k})\rangle$, namely $\langle u_i(\mathbf{k})|\mathcal{T}|u_j(\mathbf{k})\rangle$. This yields a matrix which is antisymmetric. Hence, we have

$$\langle u_i(\mathbf{k})|\mathcal{T}|u_j(\mathbf{k})\rangle = \epsilon_{ij}P(\mathbf{k}), \qquad (5.10.1)$$

where ϵ_{ij} is the Levi-Civita symbol and $P(\mathbf{k})$ is the Pfaffian of the matrix defined as

$$P(\mathbf{k}) = \mathrm{Pf}\left[\langle u_i(\mathbf{k})|\Theta|u_j(\mathbf{k})\rangle\right]. \qquad (5.10.2)$$

For a 2×2 antisymmetric matrix A_{ij}, the Pfaffian picks up the off-diagonal component. Now the absolute value of this Pfaffian is unity in the even subspace, while it is zero in the odd subspace. Hence, we dissect the BZ into two halves, such that the points $\mathbf{k}_1$ and $\mathbf{k}_2$ lie in different halves. Thus, the $\mathbb{Z}_2$ index can be computed via performing the integral

$$\mathbb{Z}_2 = \frac{1}{2\pi i}\oint_C d\mathbf{k}\cdot\boldsymbol{\nabla}\log\left(P(\mathbf{k})+i\delta\right), \qquad (5.10.3)$$

where δ is the convergence factor and the contour C is the circumference of the halved BZ discussed above.

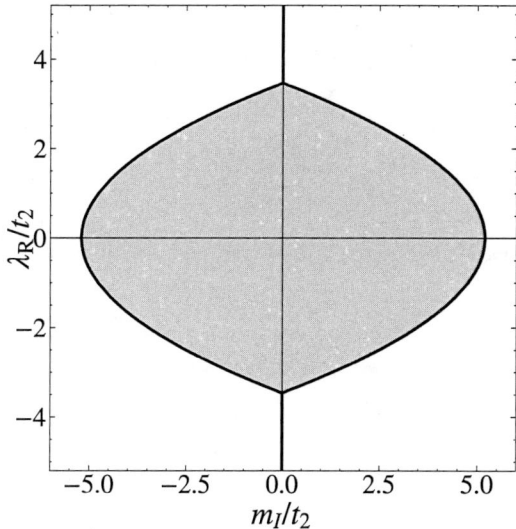

FIGURE 5.12. The phase diagram of the Kane–Mele model. The shaded region inside the closed lobe corresponds to $\mathbb{Z}_2 = 1$, which denotes a quantum Hall insulator. The unshaded region outside the lobe refers to a trivial insulator with $\mathbb{Z}_2 = 0$.

The variation of the $\mathbb{Z}_2$ index is shown in the parameter plane defined by the Rashba coupling, λ_R and the Semenoff mass, m_I (both scaled by the NNN hopping t_2), which is shown to have a value 1 in the shaded region, and vanishes outside in Fig. 5.12. The region with non-zero $\mathbb{Z}_2$ invariant will host spin filtered chiral edge modes, and will denote a QSH insulator, and the region outside denotes a trivial insulator. Fig. 5.12 denotes the phase diagram of a QSH insulator, in the same spirit as Fig. 5.6 denotes the phase diagram for a Chern insulator. To remind the readers, the Chern number vanishes here due to the presence of the time reversal symmetry, thereby ruling out the possibility of a quantum Hall-like state. However, a new topological phase emerges, known as the QSH phase. We describe a method in brief to compute the $\mathbb{Z}_2$ invariant introduced by Fu and Kane [18] in the appendix.

5.11 Spin Hall effect

To shed light on possible applications, we give a brief description of the spin Hall effect (SHE), and cursorily on the subject of spintronics. The SHE is the generation of the spin current perpendicular to the applied charge current. It leads to the accumulation of spins of opposite kinds at the edges of the sample. The spin selection can be facilitated by a strong SOC. Strong SOC may be intrinsic to doped semiconductors. The proposal has triggered intense investigation of the phenomenon, and the lead has been taken by the first observation of SHE in *n* doped semiconductors, [25] and 2D hole gases [26]. Both the experiments measure directly the spin accumulation induced at the edges of the sample through different optical techniques. However, more quantitative and accurate estimates are obtained by measuring the Hall angle. An excellent review of the family of the SHEs, comprising of SHE (discussed briefly earlier), inverse SHE, in which a pure spin current generates a charge current, and even an anomalous Hall effect (AHE), in which a charge current generates a polarized transverse charge current in a ferromagnetic material is given by J. Sinova *et al.* [27].

Let us briefly look at the early experimental determination of SHE. In the magneto-optical detection of SHE in thin films of platinum (Pt) and tungsten (W), the generation of spin currents from charge currents in the presence of SOC is shown via Kerr rotation spectroscopy [23] in Ref. [24]. In Fig. 5.13, the measured Kerr rotation angles, θ_K, as a function of line scan (*y*) are shown for Pt and W samples of 10nm–15nm width at applied current densities in a particular range of values (see Ref. [24] for details). The Kerr rotation signals are discernible for both the samples as shown in Figs. 5.13(a) and (b). Apart from the irregularities due to reflection from the edges, θ_K remains approximately constant with mutually opposite signs for Pt and W. In Figs. 5.13(c) and (d), θ_K as average values of the line scan are shown to vary linearly with the current density *j*. The results are conclusive in demonstrating SHE using Kerr rotation spectroscopy studies via spatial evolution of the spin dynamics. Possibly the corresponding data for extrinsic SHE (in the presence of an external field) should have a stronger signal, but the above results still confirm the possibility of realizing intrinsic SHE in experiments.

5.11.1   Spin current

Measuring the spin current is central to the study of SHE and hence to the emerging field of spintronics. The spin current has to be contrasted with the charge current. The charge current density, $j_{el}(\mathbf{r}, t)$, is given by

$$j_{el}(\mathbf{r}, t) = \text{Re}\left[\psi^\dagger(\mathbf{r}, t)(e\mathbf{v})\psi(\mathbf{r}, t)\right],\qquad(5.11.1.1)$$

which further obeys a continuity equation of the form,

$$\frac{d\rho^{el}}{dt} + \nabla \cdot j_{el} = 0,\qquad(5.11.1.2)$$

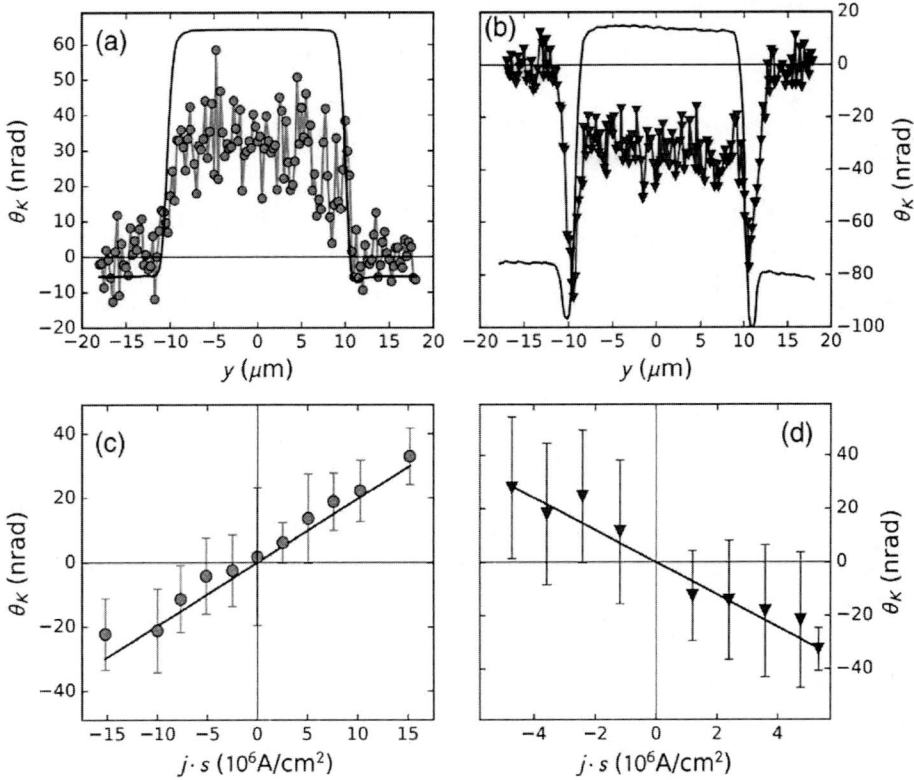

FIGURE 5.13. Kerr rotation spectroscopy studies (taken from Ref. [24]) are shown. In (a) and (b), the Kerr rotation angles θ_K as a function of y-scan (in μm) are shown for certain values of the current density, j (j in $10^6 A/cm^2$) corresponding to thin films of Pt and W respectively. (c) and (d) show linear variation of θ_K with j for Pt and W. The slopes are opposite in the two cases.

where $\mathbf{v}$ is the velocity of the electrons (charges), and $\rho^{el}(\mathbf{r}, t) = e\psi^\dagger(\mathbf{r}, t)\psi(\mathbf{r}, t)$ is the charge density. The continuity equation in Eq. (5.11.1.1) is the consequence of the invariance of charge. However, in the case of spin current, there is an ambiguity that arises from the fact that the spin is not an invariant quantity in spin transport owing to the presence of the SOC [28]. Usually, the spin current is defined as $\langle \mathbf{v.s} \rangle$ which is a non-conserved quantity. However, in the classical sense, just as the charge current density, the spin current density, $\mathbf{j}_s$ (the subscript s refers to the spin) can be written as

$$\mathbf{j}_s = \mathrm{Re}\left[\psi^\dagger(\mathbf{r}, t)(\mathbf{v.s})\psi(\mathbf{r}, t)\right]. \tag{5.11.1.3}$$

For a generic Hamiltonian with SOC, it can be shown that

$$\mathrm{Re}(\psi^\dagger v_\alpha s_\beta \psi) = \mathrm{Re}(\psi^\dagger s_\alpha v_\beta \psi),$$

where α and β refer to the components in an orthogonal coordinate system. From $\mathbf{j}_s(\mathbf{r}, t)$ one can get the total spin current using

$$I_{s\alpha}(t) = \int dA\,\hat{\alpha}.\mathbf{j}_s(\mathbf{r}, t) = \int dA \left[\psi^\dagger(\mathbf{r}, t)\frac{1}{2}(\mathbf{v}.\mathbf{s} + \mathbf{s}.\mathbf{v})\psi(\mathbf{r}, t) \right] \qquad (5.11.1.4)$$

where dA is the elemental area and $\hat{\mathbf{f}}\hat{\mathbf{f}}$ denotes a certain direction (that is, $\hat{\alpha} = (\hat{\mathbf{x}}, \hat{\mathbf{y}}, \hat{\mathbf{z}})$).

This clearly tells us that the spin current density operator, $\mathbf{j}_s$, is an anticommutator of $\{s_\alpha, v_\beta\}$ multiplied by a factor of $1/2$. In terms of the Pauli matrices,

$$j_s^{\alpha\beta} = \frac{1}{4}\{s_\alpha, v_\beta\} \qquad (\hbar = 1). \qquad (5.11.1.5)$$

Quite strikingly, unlike the charge current which is odd under time reversal, the spin current is invariant under time reversal operation. However, Ohm's law ($\mathbf{j} = \sigma\mathbf{E}$) holds for both $\mathbf{j}_{el}$ and $\mathbf{j}_s$. Since the electric field, $\mathbf{E}$ is even under time reversal, the charge conductivity, σ_{el}, is odd, while the spin conductivity, σ_s, is even.

For concreteness, let us specialize on a particular case, where we choose $\alpha = z$ and $\beta = y$. The y-component of the velocity is obtained from the Hamilton's equation of motion,

$$v_y = \frac{\partial \mathcal{H}}{\partial p_y}. \qquad (5.11.1.6)$$

Considering a Hamiltonian for a 2D electron gas with RSOC,

$$\mathcal{H} = \frac{p^2}{2m} - \lambda_R\boldsymbol{\sigma}.(\hat{\mathbf{z}} \times \mathbf{p}) \qquad \hbar = 1, \qquad (5.11.1.7)$$

which yields

$$v_y = \frac{p_y}{m} + \lambda_R\sigma_x. \qquad (5.11.1.8)$$

Finally, the spin current density assumes a form,

$$j_s^{yz} = \frac{1}{2m}\sigma_z p_y. \qquad (5.11.1.9)$$

Just as an electric current induces a magnetic field, a pure spin current induces an electric field. The magnetic moment due to the spin of the electron generates a current [29]. This can be understood as follows. Since the motion of a single magnetic moment is equivalent to an electric dipole which creates an electric field in its vicinity, there will be an electric field due to the motion of a magnetic moment. An estimation of the electric field can be made as follows. Consider two equal and opposite magnetic charges $\pm q_m$ separated by a small distance d moving in opposite directions. Such *moving* magnetic dipoles whose magnetic moments are given by $\mathbf{m} = (q_m d)\hat{\mathbf{r}}$ ($\hat{\mathbf{r}}$ denotes the polarization direction) constitute a spin current. Each member of the group, that is, a single magnetic moment will generate a magnetic field. This magnetic field in turn generates an electric field, which is given by

$$\mathbf{E} \sim \frac{\mu}{4\pi} \int dV j_s \times \frac{1}{R^3} \left(\hat{\mathbf{r}} - \frac{3\mathbf{R}(\mathbf{R}.\hat{\mathbf{r}})}{R^2} \right), \qquad (5.11.1.10)$$

where dV is an elemental volume. This electric field is quite tiny in magnitude, yet can produce measurable effects [30].

Just as a current carrying wire experiences a force in a magnetic field ($\sim \mathbf{j} \times \mathbf{B}$), a spin current experiences a force $\sim \mathbf{j}_s \times \mathbf{E}$. In spite being small, it is able to control the motion of a spin, including zitterbewegung (jittery motion) of the Dirac electrons. Thus, in semiconductors, where the SOC can be fairly strong, or can even be enhanced by external means, electrons with opposite spins are deflected along the opposite edges of the sample. Thus, a spin unpolarized (paramagnetic) system can yield a pure spin current perpendicular to the direction of the electric field.

Over the last decade and a half, studies concerning the spin current and its application to spintronics in terms of efficiently generating, manipulating and detecting the spin accumulation phenomena have received a plethora of attention. Some progress has also occurred from the device fabrication perspective, via techniques, such as spin injection. A major advantage in dealing with the spin current lies in the non-dissipative (or very less dissipation) nature which arises owing to the time reversal invariance of the spin current. This property is in direct contrast with that of the charge current. A simple way to understand the role of time reversal invariance in the phenomenon of dissipation that can be demonstrated with the aid of a damped harmonic oscillator, whose Hamiltonian may be written as

$$\mathcal{H} = \frac{p^2}{2m} + \frac{1}{2}kx^2 + \alpha \dot{x},$$

where the $\alpha \dot{x}$ denotes the damping and breaks the time reversal symmetry. Without this term, the time reversal invariance holds and the scenario is non-dissipative. Thus, a time reversal invariant system presents a non-dissipative scenario, which is precisely the main advantage for the spin transport phenomena.

5A Appendix

In the appendix, we briefly sketch the methods of finding the Chern number ($\mathbb{Z}$ invariant) following Fukui's method and the $\mathbb{Z}$ invariant using the Fu-Kane method.

5A.1 Chern number using Fukui's method

We consider a 2D system, where the BZ is defined between zero and a certain k_{max}^x along the x direction, and zero and k_{max}^y along the y direction (see Fig. 5A.1). Next, the BZ is discretized into N points along both directions. Hence, each point in the BZ is given by

$$k_l = (k_l^x, k_l^y) = \left(l\frac{k_{max}^x}{N}, l\frac{k_{max}^y}{N} \right). \tag{5.12.1.1}$$

Now we assume that the wave function $|\Psi\rangle$ is periodic on the lattice, that is,

$$|\Psi(k_l + N_\mu \hat{\mu})\rangle = |\psi(k_l)\rangle,$$

where $\hat{\mu}$ is the vector in a certain μ direction. We further define a link variable by

$$U_\mu(k_l) = \frac{\langle \Psi(k_l)|\Psi(k_l + \Delta_\mu)\rangle}{|\langle \Psi(k_l)|\Psi(k_l + \Delta_\mu)\rangle|} \tag{5.12.1.2}$$

with Δ_μ being the width in the discretized momentum space along the μ direction. In other words, $\Delta_\mu = k_{max}^\mu / N$. The link variables are well defined as long as the inner product in the denominator is non-zero.

Next, the lattice field strength, $F_{xy}(k_l)$, at a given value of momentum is calculated using the form,

$$F_{xy}(k_l) = \ln \frac{U_x(k_l)U_y(k_l + \Delta_x)}{U_x(k_l + \Delta_y)U_x(k_l)}. \tag{5.12.1.3}$$

Finally, the Chern number can be calculated by summing the field strength over all the discretized points in the BZ, which is given by

$$C = \frac{1}{2\pi i} \sum_l F_{xy}(k_l). \tag{5.12.1.4}$$

The above method is easily implementable in the numeric computation of the Chern number.

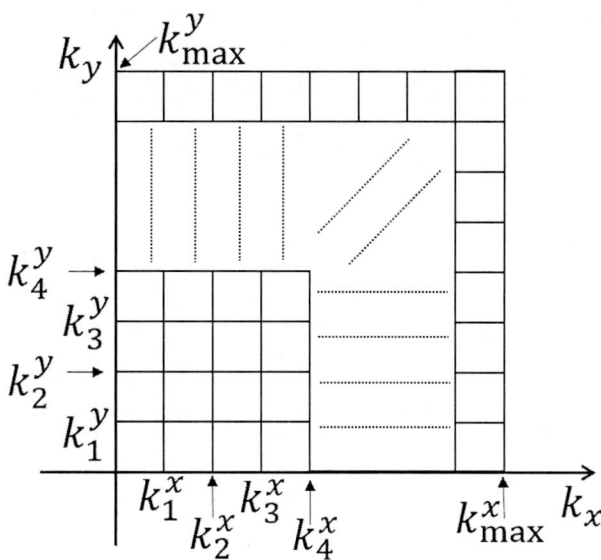

FIGURE 5A.1. The discretized Brillouin zone is shown.

5A.2 $\mathbb{Z}_2$ invariant: Fu and Kane method

It is fairly well known that if the time reversal invariance is not broken, then the Chern number[5] vanishes. However, in the presence of SOC, a system can still be topologically non-trivial with a non-zero $\mathbb{Z}_2$ invariant. The non-triviality shows up via fully gapped bulk states, while gapless edge modes continue to exist.

In fact, there are number of methods proposed to compute the $\mathbb{Z}_2$ invariant, and one of the conventional ones is to integrate the Berry connection over half of the BZ. However, this method necessitates fixing the gauge which is a challenging task in numeric computation. In the case of a centrosymmetric crystal, the task is simplified via consideration of the time reversal invariance momenta (TRIM) points where the parity eigenvalues of the occupied electronic states are required to be computed at these TRIM points.

A computationally efficient method is demonstrated by Fu and Kane [18] which relies on the localized Wannier functions (WFs),[6] instead[7] of the extended Bloch functions. Using the evolution of these WFs around closed loops in the BZ, one can study the adiabatic evolution of the occupied bands. The WFs are

---

[5]Hall angle is the angle that the resultant of the applied electric field vector (E_x) and the Hall field (E_H) makes with E_x.

[6]Also known as the first Chern number.

[7]Localized functions are anyway suited for the description of insulating phases.

also helpful for the calculation of physical properties, such as electric polarization. According to recent studies of the theory of polarization, the polarizability of the electrons involves the sum of the Berry phase of the valence band. A parallel idea relates the Berry phase of the valence bands to *Wannier charge centres* via a Fourier transformation of the Bloch functions. This facilitates computation of the ferroelectric polarization using Wannier interpolation of the Bloch bands. The power of the formalism allows generalization of the centrosymmetric crystalline materials to the non-centrosymmetric variants.

Let us define the concept of WCCs. To begin with, the WF can be written as

$$|\mathbf{R}, n\rangle = \sum_{\mathbf{R}} e^{-i\mathbf{k}.\mathbf{R}} |\psi_{n,\mathbf{k}}\rangle, \qquad (5.12.2.1)$$

where $|\psi_{n,\mathbf{k}}\rangle$ refers to the Bloch wave functions.

$$|\psi_{n,\mathbf{k}}\rangle = e^{i\mathbf{k}.\mathbf{r}} |u_{n,\mathbf{k}}\rangle. \qquad (5.12.2.2)$$

Here, $|u_{n,\mathbf{k}}\rangle$ represents a periodic function with a band index n that agrees with the periodicity of the Hamiltonian. The WFs are a real space representation of well localized orbitals that span the same Hilbert space as the Bloch wave functions. The WFs are not unique, in the sense that the gauge freedom in choosing the representatives of the Bloch wave functions in the occupied space,

$$|\tilde{u}_{n,\mathbf{k}}\rangle = \sum_{m} U_{mn}(\mathbf{k}) |u_{m,\mathbf{k}}\rangle,$$

make them gauge dependent. It is to be noted that in 1D, there always exists a unique gauge that causes the WFs to be maximally localized. In 2D and 3D, however, it is impossible to maximally localize the WFs along all the directions.

Thus, Wannier charge centres (WCCs) are basically the expectation value of the position operators with respect to the maximally localized WFs. This corresponds to the centre of charge for a particular unit cell. For 2D systems, the concept of hybrid WFs is introduced. The hybrid WFs are defined as

$$|R_x, k_y, k_z\rangle = \frac{1}{2\pi} \int dk_x e^{-iR_x k_x} |\psi_{n,\mathbf{k}}\rangle. \qquad (5.12.2.3)$$

Here, the WF is localized in the x direction, while it is periodic along the y and z directions. This forces the wave function to mimic a 1D system, even though the actual system is in 2D. In 3D, k_y and k_z can be treated as external parameters. For such a wave function, the expectation value of the position operator is calculated along the x-direction. This expectation value, however, is now a function of the variables k_y and k_z. The evolution of this expectation value as a function of k_y is called the Wannier charge centre evolution (referred to as the WCC evolution). WCC contains necessary topological information about the system.

Here, we sketch the Fu-Kane method in brief. Let us consider a crystalline lattice. The lattice constant is taken as $a = 1$ without any loss of generality. Also consider a length $L = N_c$ with periodic boundary conditions and $2N$ occupied bands. As explained earlier, one can use the WCCs to describe the topological properties of the system.

According to the definition, the charge centre corresponding to a band index n is given by

$$\mathbf{r}_n = \langle 0, n | \mathbf{r} | 0, n \rangle. \tag{5.12.2.4}$$

Further, the extent of the WF around the centre is obtained as

$$\langle r_n^2 \rangle - \langle \mathbf{r} \rangle^2 = \langle 0, n | \mathbf{r} | 0, n \rangle^2 - \mathbf{r}^2. \tag{5.12.2.5}$$

Now we shall define polarization as the sum over all the bands of the centre of charge corresponding to the Wannier states associated with $\mathbf{R} = 0$, which may be written as

$$P = \sum_n \langle 0, n | r | 0, n \rangle = \sum_n \frac{i}{2\pi} \int_{-\pi}^{\pi} \langle u_{n,\mathbf{k}} | \nabla_k | u_{n,\mathbf{k}} \rangle = \sum_n C_n, \tag{5.12.2.6}$$

where C_n is the Chern number of the nth band. The charge polarization is only defined up to a lattice constant. Consider a parameter t (called the pumping parameter) [19], such that

$$\mathcal{H}(t + T) = \mathcal{H}(t),$$

where $\mathcal{H}$ and T denote the Hamiltonian of the system and the pumping period respectively. If one changes $\mathcal{H}(t)$ adiabatically from an initial value t_1 to a final value t_2 and makes sure that the Bloch functions $|u_{n,\mathbf{k}} >$ are continuously defined during the process, the change in the charge polarization is given by

$$P[t_1] - P[t_2] = \frac{1}{2\pi} \left[\oint_{c_2} dk \mathcal{A}(t, \mathbf{k}) - \oint_{c_1} dk \mathcal{A}(t, \mathbf{k}) \right], \tag{5.12.2.7}$$

where $c_{1(2)}$ is the loop for $\mathbf{k} = -\pi$ to π for a fixed $t = t_{1(2)}$. Here,

$$\mathcal{A}(t, \mathbf{k}) = i \sum_n \langle u_{n,\mathbf{k}}(t) | \nabla_{\mathbf{k}} | u_{n,\mathbf{k}}(t) \rangle$$

is the Berry connection. Now, the following expression for the Berry curvature used to rewrite the change in charge polarization,

$$\mathcal{F} = i \sum_n [\langle \nabla_t u_{n,\mathbf{k}}(t) | \nabla_{\mathbf{k}} u_{n,\mathbf{k}}(t) \rangle - \langle \nabla_{\mathbf{k}} u_{n,\mathbf{k}}(t) | \nabla_t u_{n,\mathbf{k}}(t) \rangle]. \qquad (5.12.2.8)$$

Therefore, the change in polarization reads as

$$P[t_1] - P[t_2] = \frac{1}{2\pi} \int_{\tau_{1,2}} dt \, d\mathbf{k} \mathcal{F}(t, \mathbf{k}). \qquad (5.12.2.9)$$

For a whole period T, the quantity in the above equation, $P(T) - P(0)$, is the integral over an entire torus in the space spanned by t and $\mathbf{k}$ which is an integer known as the first Chern number. It should be noted that the system is time-reversal invariant and hence the total Chern number vanishes.

Now, to calculate the $\mathbb{Z}_2$ invariant, Fu and Kane split the total charge polarization P in two parts because of the presence of Kramer's pairs. Let's say we have $2N$ bands where each pair can be labelled as $\alpha = 1, ..., N$ and the bands corresponding to each α are assigned with an index s which can take values, $s = \mathrm{I}, \mathrm{II}$ as shown in Fig. 5A.2. Now, the advantage of having Kramer's pairs is that the time

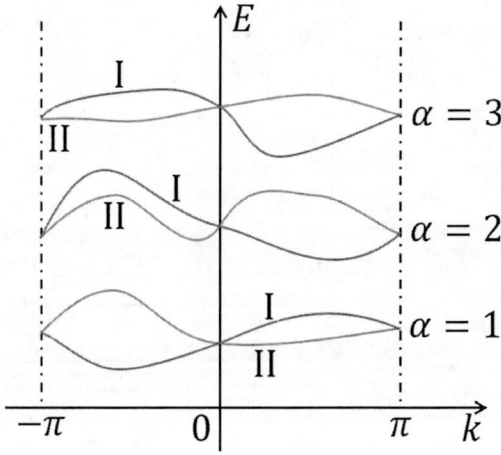

FIGURE 5A.2.   Kramer's pairs of bands of a system. For $\alpha = 1$, the eigenfunction at $k = \pi$ (for $s = \mathrm{I}$) can be obtained by operating the time reversal symmetry operator on the eigenfunction at $k = -\pi$ (for $s = \mathrm{II}$) up to a phase factor.

reversed Bloch wave at a given value of $\mathbf{k}$ is related to the Bloch wave at $-\mathbf{k}$ up to a phase factor. This may be expressed as

$$|u^{\mathrm{I}}_{-\mathbf{k},\alpha}\rangle = -e^{i\chi_{\mathbf{k},\alpha}} \mathcal{T} |u^{\mathrm{II}}_{\mathbf{k},\alpha}\rangle \qquad (5.12.2.10)$$

$$|u^{\mathrm{II}}_{-\mathbf{k},\alpha}\rangle = e^{i\chi_{-\mathbf{k},\alpha}} \mathcal{T} |u^{\mathrm{I}}_{\mathbf{k},\alpha}\rangle, \qquad (5.12.2.11)$$

where $\mathcal{T}$ is the time-reversal operator with the property $\mathcal{T}^2 = -1$. The partial polarization associated with the splitting of $s = \mathrm{I}, \mathrm{II}$ is given by

$$P^s = \frac{1}{2\pi} \int_{-\pi}^{\pi} d\mathbf{k} \mathcal{A}^s(\mathbf{k}), \qquad (5.12.2.12)$$

where

$$\mathcal{A}^s(\mathbf{k}) = i \sum_{\alpha} \langle u^s_{\mathbf{k},\alpha} | \nabla_{\mathbf{k}} | u^s_{\mathbf{k},\alpha} \rangle. \qquad (5.12.2.13)$$

The $\mathbb{Z}_2$ invariant is defined by the difference of the partial polarizations, namely the time reversal polarization, $P_\theta = P^{\mathrm{I}} - P^{\mathrm{II}}$. Since the system is time-reversal invariant, only the integration over half a period is needed. The $\mathbb{Z}_2$ invariant for such time-reversal invariant systems can be defined as

$$\nu = [P_\theta(T/2) - P_\theta(0)] \bmod 2. \qquad (5.12.2.14)$$

For time reversal symmetric systems possessing an additional inversion symmetry, the Fu-Kane method prescribes calculation of the $\mathbb{Z}_2$ invariant by utilizing the inversion eigenvalues of the occupied band subspace at the TRIMs. Given that the occupied subspace consists of N bands, the $\mathbb{Z}_2$ invariant can be calculated as

$$(-1)^\nu = \prod_{n \in \frac{N}{2}} \prod_{\mathbf{k} \in \mathrm{TRIMs}} \xi_{n,\mathbf{k}}, \qquad (5.12.2.15)$$

where $\xi_{n,\mathbf{k}} = \pm 1$ corresponds to the inversion eigenvalues, and ν denotes the required $\mathbb{Z}_2$ invariant.

6

Fractional Quantum Hall Effect

6.1 Introduction

The fractional quantum Hall effect (FQHE) was discovered by Tsui, Stormer and Gossard in 1982 at Bell Labs. They observed that at very high magnetic fields, a 2DEG shows fractional quantization of the Hall conductance. In particular, they got a quantized Hall plateau of magnitude $\rho_{xy} = \frac{3h}{e^2}$, which is accompanied by the vanishing of the longitudinal conductivity, ρ_{xx}, at low temperature ($T < 5$ K) in GaAs and AlGaAs samples. As opposed to the integer quantum Hall effect (IQHE), where an integer number of Landau levels (LLs) are occupied, here in FQHE the LLs are partially occupied. If one makes the magnetic field large enough, the lowest Landau level (LLL) will be partially filled. What we can expect is that the system will form some kind of a lattice, for example, a Wigner crystal or a charge density wave. Thus, it naively seems to be reasonable that the system would like to minimize its potential energy, since there is no kinetic energy left in the system corresponding to the LLL, and only a trivial zero point energy is present in the system. Thus, the ions tend to stay away from each other and form something similar to a crystal lattice. However, surprisingly that does not happen, and instead the system becomes an incompressible quantum liquid, which has gaps in the energy spectrum at filling $1/m$ (m: odd, or a rational fraction of the form n/m). So it is inevitable that the system minimizes its energy by having gaps at fractional values of filling. The reason is that, owing to the presence of a large number of electrons (macroscopically degenerate in any of the LLs), a many-body interaction is induced, which in fact makes the excitations above this incompressible ground state to be fractional. So in essence, the Hall current carries a fractional charge.

In fact, the excitations are called abelian anyons, which are neither fermions nor bosons; if they are taken twice around a complete circle, they will pick up a phase that is either 0 or π. The phenomenology is put forward by J. K. Jain [92] in terms of composite fermions (CFs), which we shall describe later.

One important thing that becomes apparent from the preceding discussion is that FQHE is impossible to explain by invoking the interaction between the electrons that eventually split the degeneracy of these enormously degenerate LLs, leading to the opening of a gap. The gap is in principle similar to the cyclotron gap, $\hbar\omega_c$, seen earlier in IQHE. However, this introduces another energy scale, leading to a hierarchy of energy scales, for example, the Coulomb energy, namely $e^2/\epsilon a_0$ (the length scale in the denominator is the Bohr radius) and the cyclotron energy (due to the applied magnetic field). Thus, the kinetic degrees of freedom of the electrons are frozen and hence get eliminated from the problem. As said earlier, while the expectation is that of a trivial insulating state, but at fractional fillings (less than 1), something more interesting happens.

Now consider the effect of the Coulomb interaction between the electrons,

$$V = \frac{e^2}{|\mathbf{r_i} - \mathbf{r_j}|}. \tag{6.1.1}$$

This interaction should lift the degeneracy of the ground states. A degenerate perturbation theory seems to be an answer to this. However, degenerate perturbation theory problems are only solvable if the degree of degeneracy is such that it is analytically or numerically tractable. But here we are stuck with extraordinary large matrices to diagonalize. Even numerically, with the best computational resources available nowadays, one cannot diagonalize for more than a few particles.

In the absence of interactions, we have sharply defined LLs. In the presence of interactions, the degeneracy goes away and the LL broadens, resulting in a spectrum of states width $\sim E_{Coulomb}$. Now the experimental data would be nicely explained if a (tiny) gap exists for a filling fraction $\nu = \frac{1}{3}$ ($\nu = \frac{1}{3}$ is just an example), and the gap can exist at any of the rational fractions for which FQHE is seen. Of course, the more prominent the plateau in the FQHE plot (shown in Fig. 1.4), the larger is the gap. So we include some disorder that introduces localized states within the gap, which then gives rise to plateaus in ρ_{xy} (and $\rho_{xy} = 0$). The whole description requires a hierarchy of energy scales, namely

$$\hbar\omega_B \gg E_{Coulomb} \gg V_{disorder}. \tag{6.1.2}$$

6.2 Electrons in the symmetric gauge

Based on our earlier discussion, it is obvious that owing to the presence of a very large number of electrons in a single LL (each level has practically infinite occupancy), the interaction effects cannot be neglected. When we talk about interaction between electrons, it is most natural to think about Coulomb interaction. The electronic orbitals in this case have radial forms. A symmetric (or a circular) gauge is most suited for this description, as opposed to the Landau gauge used earlier to describe IQHE. The Laughlin wavefunction (discussed later) written in terms of a complex number, z ($= x + iy$ or $\rho e^{i\phi}$), which describes the complex coordinate of the electrons, is also compatible with the symmetric gauge.

The vector potential $\mathbf{A}$ in the mixed gauge can be written as (again yields $\mathbf{B} = B\hat{z}$)

$$\mathbf{A} = -\frac{1}{2}(\mathbf{r} \times \mathbf{B}) = \frac{1}{2}(-By\hat{x} + Bx\hat{y}). \tag{6.2.1}$$

The choice of the symmetric gauge breaks the translation symmetry in both the x and y directions. However, it preserves rotational symmetry about the origin. This of course means that angular momentum is a good quantum number. This is the most convenient gauge to study FQHE. Further, we can write down the non-canonical momentum as

$$\pi = \mathbf{p} + e\mathbf{A} = m\dot{\mathbf{r}}. \tag{6.2.2}$$

This is gauge invariant but non-canonical. One can use this to form the raising and lowering operators:

$$a = \frac{1}{\sqrt{\frac{eBh}{\pi}}}(\pi_x - i\pi_y) = \sqrt{\frac{\pi}{eBh}}(\pi_x - i\pi_y) \tag{6.2.3}$$

$$a^\dagger = \sqrt{\frac{\pi}{heB}}(\pi_x + i\pi_y),$$

where $2e\hbar B = 2e\frac{h}{2\pi}B = \frac{eBh}{\pi}$. Now,

$$[a, a^\dagger] = aa^\dagger - a^\dagger a = \left(\frac{\pi}{heB}\right)(\pi_x - i\pi_y)(\pi_x + i\pi_y) \tag{6.2.4}$$
$$- (\pi_x + i\pi_y)(\pi_x - i\pi_y)$$
$$= \left(\frac{\pi}{eBh}\right)[\pi_x^2 + i\pi_x\pi_y - i\pi_y\pi_x + \pi_y^2 - \pi_x^2 + i\pi_x\pi_y - i\pi_y\pi_x - \pi_y^2]$$

$$= \left(\frac{2\pi i}{eBh}\right)[\pi_x\pi_y - \pi_y\pi_x]$$

$$= \left(\frac{i}{eBh}\right)[\pi_x, \pi_y].$$

π is not the canonical momenta in the sense, $[x_i, \pi_j] \neq \delta_{ij}$ and $[\pi_i, \pi_j] = \delta_{ij}$. However, they are gauge invariant. The numerical value of π does not depend upon the choice of gauge. This can be proved by the commutation relations

$$\{p_i + eA_i, p_j + eA_j\} = -e\left(\frac{\partial A_j}{\partial x^i} - \frac{\partial A_i}{\partial x^j}\right) = -e\varepsilon_{ijk}B_k. \tag{6.2.5}$$

Thus, $[a, a^\dagger] = (\frac{i}{eBh})(-ie\hbar B) = 1$.

To explore whether the LLs yield the expected degeneracy, we can introduce another momentum variable, namely

$$\tilde{\pi} = \mathbf{p} - e\mathbf{A}.$$

$\tilde{\pi}$ is not gauge invariant and depends on the choice of the gauge potential. The vector potential, $\mathbf{A}$, enjoys the gauge freedom, that is,

$$\mathbf{A}' = \mathbf{A} - \nabla\chi,$$

where χ is an arbitrary scalar. This yields

$$\tilde{\pi}' = \mathbf{p}' - e(\mathbf{A}' - \nabla\chi) = \mathbf{p}' - e\mathbf{A}' + e\nabla\chi.$$

The commutation relation for $\tilde{\pi}'$ is

$$[\tilde{\pi}'_x, \tilde{\pi}'_y] = ie\hbar B, \tag{6.2.6}$$

which differ only by a sign with respect to the π momenta. This is also an advantage of these new momenta in that they obey

$$[\tilde{\pi}_i, \tilde{\pi}_j] = 0.$$

Finally, the Hamiltonian is written in terms of the a and $a^\dagger$ operators as

$$\mathcal{H} = \hbar\omega_B\left(a^\dagger a + \frac{1}{2}\right) = \hbar\omega_B\left(n + \frac{1}{2}\right), \tag{6.2.7}$$

where $n = 0, 1, \ldots$ and denote the indices of the LLs.

Now as a matter of exercise, in order to explore the degeneracy of the LLs, we introduce a second pair of raising and lowering operators, namely

$$b = \frac{1}{\sqrt{2e\hbar B}}(\tilde{\pi}_x + i\tilde{\pi}_y) \tag{6.2.8}$$

$$b^\dagger = \frac{1}{\sqrt{2e\hbar B}}(\tilde{\pi}_x - i\tilde{\pi}_y). \tag{6.2.9}$$

They too obey $[b, b^\dagger] = 1$. These b, $b^\dagger$ will yield the degeneracy of the LLs as shown in the following discussion. Thus, a general state in the Hilbert space $|n, m\rangle$ is defined by

$$|n, m\rangle = \frac{(a^\dagger)^n (b^\dagger)^m}{\sqrt{n!m!}} |0, 0\rangle, \tag{6.2.10}$$

where $a |0, 0\rangle = b |0, 0\rangle = 0$ and $\mathcal{H} = \frac{1}{2m}\boldsymbol{\pi} \cdot \boldsymbol{\pi} = \frac{1}{2m}(\mathbf{p} + e\mathbf{A})^2$.

Let's now construct the wavefunction in the symmetric gauge. We are going to focus on the LLL, $n = 0$, as it is of primary interest for discussing FQHE. The trick is to convert the definition of a into a differential equation

$$a = \frac{1}{\sqrt{2e\hbar B}}(\pi_x - i\pi_y) \tag{6.2.11}$$

$$= \frac{1}{\sqrt{2e\hbar B}}[p_x - ip_y + e(A_x - iA_y)]$$

$$= \frac{1}{\sqrt{2e\hbar B}}\left[-i\hbar\left(\frac{\partial}{\partial x} - i\frac{\partial}{\partial y}\right) + \frac{eB}{2}(-y - ix)\right] \tag{6.2.12}$$

using $z = x - iy$ and $\tilde{z} = x + iy$. Remember this is not usually how we define z and z^* (or $\tilde{z}$); however, we shall stick to this definition. We also define

$$\partial = \frac{1}{2}\left(\frac{\partial}{\partial x} + i\frac{\partial}{\partial y}\right) \tag{6.2.13}$$

and

$$\tilde{\partial} = \frac{1}{2}\left(\frac{\partial}{\partial x} - i\frac{\partial}{\partial y}\right), \tag{6.2.14}$$

which obey $\partial z = \tilde{\partial}\tilde{z} = 1$ and $\partial\tilde{z} = \tilde{\partial}z = 0$.

$$\partial z = \frac{1}{2}(\frac{\partial}{\partial x} + i\frac{\partial}{\partial y})(x - iy) = \frac{1}{2}(1 + 1) = 1.$$

So a and $a^\dagger$ in terms of the coordinates z can be written as

$$a = -i\sqrt{2}\left(l_B\tilde{\partial} + \frac{z}{4l_B}\right); \quad a^\dagger = -i\sqrt{2}\left(l_B\partial - \frac{\tilde{z}}{4l_B}\right). \qquad (6.2.15)$$

Now, the LLL is found by the one which is annihilated by this operator a.

$$a\,|0,m\rangle = 0$$

$$-i\sqrt{2}\left(l_B\tilde{\partial} + \frac{z}{4l_B}\right)|0,m\rangle = 0,$$

$|0,m\rangle$ is called $\psi_{\text{LLL}}(z,\tilde{z})$, where LLL stands for the lowest Landau level.

$$\psi_{\text{LLL},m=0} \simeq e^{-|z|^2/4l_B^2}.$$

The ground state is known to be a Gaussian for a linear Harmonic oscillator.[1]

One can construct the higher LL wavefunctions by employing $b^\dagger$ successively to the $m = 0$ state. This yields

$$\psi_{\text{LLL},m} \simeq \left(\frac{z}{l_B}\right)^m e^{-|z|^2/4l_B^2}. \qquad (6.2.16)$$

It is straightforward to ascertain that $\psi_{\text{LLL},m}$ are eigenfunctions of J_z, defined by

$$J_z = \hbar(z\partial - \tilde{z}\tilde{\partial}) \qquad (6.2.17)$$

and obey $J_z\psi_{\text{LLL},m} = m\hbar\psi_{\text{LLL},m}$.

Let us explore the degeneracy associated with the LLs $\psi_{\text{LLL},m}$, which is obtained by noting that the wavefunction with angular momentum, m, is peaked on a circular ring of radius, $r = l_B\sqrt{2m}$. The number of states in an area $A = \pi R^2$ is $\mathcal{N} = \pi R^2/\pi r^2 \simeq eBA/2\pi\hbar$, which is a result that we have seen earlier.

A quick recap of the ongoing discussion reveals that for a constant magnetic field B, the vector potential can be obtained as

$$\mathbf{A} = -\frac{1}{2}(\mathbf{r} \times \mathbf{B}) = \frac{B}{2}(-y\hat{\mathbf{x}}, x\hat{\mathbf{y}}, 0). \qquad (6.2.18)$$

---

[1] $(y + \frac{\partial}{\partial y})u(y) = 0$, or $\frac{\partial u}{u} = -ydy$. The solution is $u = e^{-y^2/2}$.

Writing down the free particle Hamiltonian $\mathcal{H} = \frac{1}{2m}(\mathbf{p} + e\mathbf{A})^2$ in the above gauge,

$$\mathcal{H} = \frac{1}{2l_B^2}\left[\left(-i\frac{\partial}{\partial x} - \frac{y}{2}\right)^2 + \left(-i\frac{\partial}{\partial y} + \frac{x}{2}\right)^2\right]. \tag{6.2.19}$$

It is convenient to introduce the complex variables z and z^* via

$$z = x - iy = re^{-i\theta}, \quad z^* = x + iy = re^{i\theta}.^2$$

The derivatives can be written as

$$\frac{\partial}{\partial x} = \frac{\partial}{\partial z} + \frac{\partial}{\partial z^*}, \quad \frac{\partial}{\partial y} = -i\left(\frac{\partial}{\partial z} - \frac{\partial}{\partial z^*}\right).$$

In terms of z and z^*, the Hamiltonian reads as

$$\mathcal{H} = \frac{1}{2l_B^2}\left[\frac{1}{4}|z|^2 - 4\frac{\partial^2}{\partial z\partial z^*} - z\frac{\partial}{\partial z} + z^*\frac{\partial}{\partial z^*}\right]. \tag{6.2.20}$$

The Hamiltonian has little similarities with that of the harmonic oscillator, especially because of the mixed second derivative and the two first derivatives. In order to solve the Hamiltonian, a set of ladder operators can be introduced, namely

$$b = \frac{1}{\sqrt{2}}\left(\frac{z^*}{2} + 2\frac{\partial}{\partial z}\right), \quad b^\dagger = \frac{1}{\sqrt{2}}\left(\frac{z}{2} - 2\frac{\partial}{\partial z^*}\right) \tag{6.2.21}$$

$$a = \frac{1}{\sqrt{2}}\left(\frac{z}{2} + 2\frac{\partial}{\partial z^*}\right), \quad a^\dagger = \frac{1}{\sqrt{2}}\left(\frac{z^*}{2} - 2\frac{\partial}{\partial z}\right).$$

These operators obey

$$[a, a^\dagger] = [b, b^\dagger] = 1 \tag{6.2.22}$$

with all other commutators vanishing. This facilitates writing the Hamiltonian in a familiar form,

$$\mathcal{H} = a^\dagger a + \frac{1}{2}. \tag{6.2.23}$$

The eigenvalue of $a^\dagger a$ denotes the LL index n.

---

[2]Somewhat non-trivial definitions of z and z^* are adopted to make sure of the analyticity of the wavefunction in z.

Let us now concentrate on the *b*-operators. They have a role to play in writing down the *z*-component of the angular momentum, namely J_z, where

$$J_z = -i\hbar \frac{\partial}{\partial \phi} = -\hbar \left(z \frac{\partial}{\partial z} - z^* \frac{\partial}{\partial z^*} \right) = a^\dagger a - b^\dagger b. \tag{6.2.24}$$

The eigenvalue of J_z is $-m\hbar$, where m takes values from $-n$ to $+n$, n being the LL index. $b^\dagger(b)$ increases (decreases) the value of m by one unit, while keeping the LL index unchanged. However, $a^\dagger(a)$ increases (decreases) n and decreases (increases) m by one unit.

Clearly, the harmonic oscillator problem here has two indices, namely n and m, which are apparent through

$$\mathcal{H}|n,m\rangle = E_n|n,m\rangle, \tag{6.2.25}$$

where

$$|n,m\rangle = \frac{(b^\dagger)^{n+m}}{\sqrt{(n+m)!}} \frac{(a^\dagger)^n}{\sqrt{n!}} |0,0\rangle,$$

where $|0,0\rangle$ denotes a Gaussian

$$|0,0\rangle = \frac{1}{\sqrt{2\pi}} e^{-|z|^2/4l_B^2}$$

and obeys

$$a|0,0\rangle = 0 = b|0,0\rangle.$$

One can generate the family of LLLs by successively operating $b^\dagger$ on $|0,0\rangle$, that is,

$$(b^\dagger)^m|0,0\rangle = |0,m\rangle = \frac{z^m e^{-|z|^2/4l_B^2}}{\sqrt{2\pi 2^m m!}}.$$

$\sqrt{2\pi 2^m m!}$ appears as the normalization constant, and z^m is a polynomial in z, where each z appears due to acting $b^\dagger$ on $|0,0\rangle$ each time.

It can be checked that $|0,m\rangle$ is an analytic function of z, since

$$\frac{\partial}{\partial z^*}|0,m\rangle = 0,$$

which is evident from the definition of $\frac{\partial}{\partial z^*} = \frac{1}{2}\left(\frac{\partial}{\partial x} - i\frac{\partial}{\partial y} \right)$ as the Cauchy–Riemann condition is satisfied.

Now getting an LL with arbitrary indices, namely $|n,m\rangle$, we need to simultaneously act $a^\dagger$ n times and $b^\dagger$ $(n+m)$ times. This is necessary because $a^\dagger$ reduces the m index by one unit each time. Thus,

$$|n,m\rangle = \frac{1}{2\pi 2^{m+2n} n!(n+m)!} \left(a^\dagger\right)^n \left(b^\dagger\right)^{n+m} e^{-|z|^2/4l_B^2} \tag{6.2.26}$$

$$= \frac{1}{2\pi 2^{m+2n} n!(n+m)!} \left(z^* - 2\frac{\partial}{\partial z^*}\right)^n \left(z - 2\frac{\partial}{\partial z}\right)^{m+n} e^{-|z|^2/4l_B^2}.$$

It has a somewhat complicated form, but at least denotes an expression that can be evaluated for a given value of n and m.

Let us more closely discuss the physics of these fractional quantum Hall states. The first approach to the fractional quantum Hall states was due to Laughlin, who described the filling fractions given by $v = 1/m$ (m is an odd integer). Since the resultant matrix was too difficult to diagonalize, Laughlin wrote down the answer in the following sense. He wrote down the wavefunction by intuition that preserves the physical properties and the symmetries of the system. To understand the Laughlin wavefunction, let us consider only two particles in the LLL. Consider a potential, $V = V(|\mathbf{r_1} - \mathbf{r_2}|)$. For such a potential, the wavefunction is an eigenstate of the angular momentum (recall the H-atom problem). In order for the angular momentum basis to be used, we need a symmetric gauge, namely

$$\mathbf{A} = -\frac{1}{2}\left(\mathbf{r} \times \mathbf{B}\right) = -\frac{By}{2}\hat{\mathbf{x}} + \frac{Bx}{2}\hat{\mathbf{y}}. \tag{6.2.27}$$

The choice of the gauge breaks the translation symmetry in both the x and the y direction; however, it preserves rotational symmetry about the origin. This means that angular momentum is a good quantum number and hence justifies using the angular momentum basis.

6.3 The lowest Landau level

The (unnormalized) single particle wave function in the LLL takes the form

$$\psi_m = z^m e^{-|z|^2/4l_B^2}, \tag{6.3.1}$$

with $z = x - iy$. These states are located on a ring of radius $r = l_B\sqrt{2m}$. The exponent m labels the angular momentum. The largest value of m for which the state falls inside the ring is given by $m = R^2/2l_B^2$, R being the corresponding value of the radius of the ring. m now denotes the total number of eigenstates in the LLL that falls inside the ring. That is, $0 \leq m \leq N_\Phi$, where $N_\Phi = \frac{A}{\Phi_0} n$ is the number of flux quantum. Hence, the degeneracy per unit area is given by $\frac{1}{2\pi l_B^2} = \frac{eB}{h} = \frac{B}{h/e}$. As long as we neglect mixing between the successive LLs, that is, when $V \ll \hbar \omega_B$, a two-particle eigenstate takes a form

$$\psi = (z_1 + z_2)^M (z_1 - z_2)^m e^{-(|z_1|^2 + |z_2|^2)/4l_B^2}, \tag{6.3.2}$$

where M is the centre of mass angular momentum and m is the relative angular momentum. M and m are non-negative integers (see discussion below), which make the pre-factor to be a polynomial in $z_1 + z_2$ and $z_1 - z_2$. Thus, a state formed out of the linear combinations of ψ_m in Eq. (6.3.2), and hence lies in the LLL. Of course, m must be odd to yield an antisymmetric wavefunction.

Now we can write down the wavefunction without explicitly solving the Schrödinger equation for any general potential of the form $V(\mathbf{r})$ (in principle, it should be an unsolvable problem). We remind ourselves that we are in the LLL that allows us to do the above form of the wavefunction. For N-particles, the many-body state can be written as

$$\psi(z_1, z_2, \cdots, z_N) = f(z_1, z_2, \cdots, z_N) e^{-\sum_{i=1}^N |z_i|^2/4l_B^2}, \tag{6.3.3}$$

where $f(z)$ is a polynomial in z_i and contains the maximum power of any z_i occurring in N_Φ. It also takes care of the statistics, that is, for interchanging $z_i \leftrightarrow z_j$, $f(z)$ picks up a negative sign.

With all the points discussed above, Laughlin made a suggestion for the ground state wavefunction [93] as

$$\psi(z_1, z_2, z_3 \cdots z_N) = \prod_{i<j} (z_i - z_j)^m e^{-\sum_{i=1}^N |z_i|^2/4l_B^2}. \tag{6.3.4}$$

So the symmetry function is fixed as $(z_i - z_j)^m$. The above form makes sense as it keeps the electrons apart and thus reduces the Coulomb interaction (or the potential energy, as discussed earlier). This is the celebrated Laughlin wavefunction for FQHE. This can be proved easily by writing

$$J_z |\psi_m\rangle = i\hbar \left(x \frac{\partial}{\partial y} - y \frac{\partial}{\partial x} \right) |\psi_m\rangle = m\hbar |\psi_m\rangle. \tag{6.3.5}$$

If m is an odd integer, for example, $1, 3, 5$, then f is an antisymmetric function.

The pre-factor vanishes of order m if two electrons happen to come close. Meanwhile, the exponential factor vanishes as i and j get far away from the origin.

The angular momentum of these states is $m\hbar$. For the LLL, there is only one state for a given value of m, and it should only be a positive value of m. The exclusion of the negative values is an artefact of the chirality created by the presence of the magnetic field.

It is worth comparing this with the scenario corresponding to the Landau gauge, which had discrete levels as the basis states. However, physical properties, such as the density of states (per unit area), should be independent of the choice of the gauge. It can be tested by taking the mod square of Eq. (6.3.1). The maximum value of this probability density occurs at $R = \sqrt{2ml_B^2}$. Thus, the area of a circle given by $\pi R^2 (= 2m\pi l_B^2)$ contains m flux quanta, which yields the familiar result of one state per LL per flux quantum piercing the quantum Hall sample.

Let us show that this wavefunction has the desired filling fraction which is an important aspect of the Laughlin state. The exponent m is not arbitrary and related to the filling fraction. For this purpose, let us focus on the wavefunction for a single particle coordinate, say, z_1. The terms that depend upon z_1 in the pre-factor can be written as

$$\prod_{i<j}(z_i - z_j)^m \sim \prod_{i=2}^{N}(z_1 - z_i)^m \tag{6.3.6}$$

$$= (z_1 - z_2)^m (z_1 - z_3)^m \cdots (z_1 - z_N)^m. \tag{6.3.7}$$

Thus, there are $m(N-1)$ powers of z_1. Hence, the maximum angular momentum of the first particle is $m(N-1)$. So the maximum extent of the wavefunction is given by the radius, $R \approx \sqrt{2mN}l_B$, where $(N-1)$ is approximated by N for large N. Correspondingly, the area over which it spans is given by $\pi R^2 \approx 2\pi m N l_B^2$.

Now, recall that the number of states in the filled LL is $AB/\Phi_0 \approx A/2\pi l_B^2$, where B/Φ_0 is the inverse of the area. Thus, putting $A = \pi R^2 = 2\pi m N l_B^2$, $l_B = \sqrt{\frac{\hbar}{eB}}$, the total number of states is given by

$$\frac{2\pi m N l_B^2}{2\pi l_B^2} = mN. \tag{6.3.8}$$

Since the total number of states is mN, the filling fraction is $\frac{1}{m}$ (m: odd) as we have discussed earlier. The exact diagonalization of the Hamiltonian matrix for a small number of particles shows that the Laughlin wavefunction is extremely accurate.

6.4 The filling fraction revisited

Let us revisit the above discussion in a slightly different language. The total angular momentum is the sum of the angular momentum of individual electrons. So the total angular momentum carried by the Laughlin state is as follows.

A typical term in the Laughlin wave function is

$$z_1^0 z_2^m z_3^{2m} \cdots z_N^{(N-1)m} e^{-\sum_{i=1}^N |z_i^2|},$$

where N denotes the number of electrons. Since the above state has angular momentum $m\hbar$ for the individual electrons, the total angular momentum, $\mathbf{J}_{tot}$, is given by

$$|\mathbf{J}_{tot}| = m\hbar \sum_{n=0}^{N-1} n = \frac{(N-1)Nm}{2}\hbar.$$

All other contributions will yield the same angular momentum.

On the other hand, the maximum angular momentum, namely $n_{max}\hbar$, that an electron can have is given by the maximum power of the variable z in the wavefunction. Here,

$$n_{max} = (N-1)m.$$

Consider the form of the LLL.

$$\psi_m = z^m e^{-|z|^2/4l_B^2}.$$

The probability of finding the electron at a given z is stated by $|\psi(z)|^2$ (or $|\psi(r)|^2$). This quantity has a sharp peak at $r = r_m$.

$$r_m = \sqrt{2m}l_B.$$

Thus, the area over which the LLs are located is given by

$$A = \pi r_{max}^2 = 2\pi n_{max} l_B^2$$
$$= 2\pi m(N-1)l_B^2. \tag{6.4.1}$$

Thus, the Laughlin state is realized at the filling fraction

$$\nu = \frac{N\Delta A}{A} = \frac{N}{m(N-1)} = \frac{1}{m}$$

for large N.

Another important feature of the Laughlin wavefunction is that the argument of the Gaussian can be written as

$$exp\left(-\sum_i z_i^2\right) = exp\left(-\sum_{i \neq j}|z_i - z_j|^2 + \sum_i z_i^2\right).$$

Writing $\bar{z}^2 = \sum_i z_i^2$, we note that $\bar{z}$ is the coordinate of the centre of mass. Thus, apart from this trivial factor in the exponent, the wavefunction depends on the relative (complex) coordinates, namely $|z_i - z_j|$, which imply that the wavefunction is uncorrelated and similar to the wavefunction seen for the integer version of the QHE.

This brings us to the most potential issue: whether the much talked about Laughlin wavefunction yields the Hall conductivity quantized as $v\frac{e^2}{h}$, with $v = \frac{1}{m}$ and m being an odd integer? Further, what does a fractional coefficient of $\frac{e^2}{h}$ exactly mean in terms of fractionalizing the unit of charge?

6.5 Fractional charge and the Hall conductivity

The existence of the fractional electron charge is surprising, as electrons are known to be indivisible objects. Also, and probably more importantly, how does the Laughlin wavefunction produce the Hall conductivity with plateaus at rational fractions? Consider the Corbino disc argument applicable to the IQHE (see Fig. 1.12). In order to introduce a connection between the Corbino disc model and the Laughlin wavefunction written above, we consider enhancing the magnetic field B in a controlled manner so as to introduce one flux quantum, Φ_0, in any region of the disc. The question is 'how does that modify the Laughlin wave function?' Recall that the pre-factor of the Gaussian in the Laughlin wavefunction is given by z^m (see Eq. 6.3.1), where the index m counts the number of flux quantum. Thus, with the above modification, 'm' increases to '$m + 1$', which makes the maximum power of z_i to increase from $m_{max} = Nm$ to $Nm + 1$. The situation to the Laughlin wavefunction is incorporated by introducing

$$\psi' = \left[\Pi_{i=1}^N z_i\right]\psi_0(z_1, z_2 \cdots z_N), \tag{6.5.1}$$

where $\psi_0(z_1, z_2 \cdots z_N)$ is the usual Laughlin state. The above wavefunction ψ' dismisses the origin to be a special point about which the original Laughlin state was centred (or equivalently, the single particle density had a peak at the origin).

Since the origin (z_0 is the origin) does not play an important role, we are allowed to write

$$\psi'(z_0) = \left[\Pi_{i=1}^{N}(z_i - z_0)\right]\psi_0(z_1, z_2 \cdots z_N), \qquad (6.5.2)$$

assuming z_0 is not within a distance l_B from the edge of the disc so that the single particle density remains uniform up to a distance l_B relative to the edge of the disc. In effect, the probability of finding an electron at the origin is missing, and its density within an area l_B^2 about the origin is reduced. In the basic language of solid state physics, a missing electron is equivalent to the appearance of a 'hole', which is what precisely happens here.

Now the magnetic field B is increased such that m flux quanta are added in the process and m holes are created. This yields the Laughlin state to assume the form

$$\psi_m = \Pi_{i=1}^{N}(z_i - z_0)^m \psi_0(z_1, z_2 \cdots z_N) \qquad (6.5.3)$$

$$= \Pi_{i=1}^{N+1}(z_i - z_0)^m exp\left(-\sum_{i=1}^{N+1}\frac{z_i^2}{4l_B^2}\right).$$

The above form precisely coincides with the case of $N + 1$ electrons and consequently corresponds to $m(N + 1)$ flux quanta. Thus, the addition of an extra electron compensates for the m added holes. Hence, the charge of the hole is

$$e_h = -\frac{e}{m}. \qquad (6.5.4)$$

This accounts for the fractional charge. But at the same time, this means that the probability of finding an electron near the origin is reduced by $1/m$. This probability helps us to reconcile the very idea of the Corbino disc, that is, transfer the 'fractional' charge from the centre to the edge of the disc.

We rerun the same argument as done for the IQHE. For the argument to be valid, the geometry of the ring is important. Here in addition to the background magnetic field **B** that threads the sample, we can thread an additional flux Φ through the centre of the ring. This Φ can affect the quantum state of the electrons.

Let us first see what this flux Φ has got to do with the Hall conductivity. Suppose we slowly increase Φ from 0 to $\Phi_0(= \frac{h}{e})$, that is, within a time $t_0 >> \frac{1}{\omega_B}$. This induces an emf around the annular region $\varepsilon = -\frac{\partial\Phi}{\partial t} = \frac{-\Phi_0}{t_0}$. The purpose of this emf is to transport 'n' electrons from the inner circumference to the outer circumference. This would result in a current in the radial direction, $I_r = -ne/t_0$. Thus, the Hall resistivity is

$$\rho_{xy} = \frac{\varepsilon}{I_r} = -\frac{\Phi_0}{t_0} \cdot \frac{t_0}{(-ne)} = \frac{h}{e^2} \cdot \frac{1}{n}. \tag{6.5.5}$$

The same arguments hold equally for the IQHE and FQHE; in the former n is an integer, while n is a fraction for the latter. In FQHE, the interpretation is as follows: as we increase the flux from Φ to Φ_0, a charge of e/m is transported from the inner circumference to the outer one when the flux is increased by Φ_0 units. The resultant Hall conductivity (or equivalently the resistivity) becomes

$$\sigma_{xy} = \frac{e^2}{h} \cdot \frac{1}{m}. \tag{6.5.6}$$

Thus, a whole electron is transferred only when the flux is increased by $m\Phi_0$ units.

Shot noise measures of the fractional plateaus indeed confirm the existence of fractional charge. In the experimental set up [97], the two energy modes at opposite edges of the Hall sample are coupled by a quantum point contact that facilitates a flow of current between the two edge channels. The random transfer of charges yields fluctuations (noise) of the current. In the weak coupling regime, the noise intensity is proportional to the backscattered current and the (fractional) charge. Thus, the intensity of the shot noise experiments detects fractional electronic charges.

6.6 Fractional Hall fluid and the plasma

The variational wavefunction corresponding to the LLL can be written as

$$\psi_m(z_1...z_N) = \Pi_{j<k}^{N}(z_j - z_k)^m e^{-\Sigma_j |z_j|^2/4l_B^2}. \tag{6.6.1}$$

The wavefunction is applicable to a filling fraction $\nu = \frac{1}{m}$ (m is an odd integer), such as $\nu = \frac{1}{3}$ (for $m = 3$). However, if only $1/m$ fractions were found in experiments, life would have been simpler. Unfortunately, many other fractions were found, including the improper ones.

Let us enumerate a few properties of the wavefunctions.

(i) It is antisymmetric with respect to swapping of coordinates of the fermions ($z_j \leftrightarrow z_k$),

$$\psi_m(z_1....z_j, z_k....z_N) = -\psi_m(z_1....z_k, z_j....z_N).$$

In order for this to be valid, the index m must be odd.

(ii) A surprising fact is revealed when the unnormalized probability density of the Laughlin states is computed. That is,

$$|\psi_m(z)|^2 = |\Pi_{j<k}^N (z_j - z_k)^m e^{-\sum_j |z_j|^2/4l_B^2}|^2.$$

The two terms (the Jastrow factor and the Gaussian) in the RHS behave differently. That is, the pre-factor (of the Gaussian) or the Jastrow factor tries to keep the fermions away, and it grows larger as they move further away. However, the exponential term shrinks as the fermions spread out. Under this competing scenario, can it ensure a uniform density?

There is an answer to the rather complicated problem, and again it is due to Laughlin, via an analogy with classical plasma, albeit the Hall fluid is at a very low temperature. The norm of the wavefunction is written as

$$|\psi_m(z)|^2 = e^{-\beta V_{\text{plasma}}}, \qquad (6.6.2)$$

where

$$V_{\text{plasma}} = 2m^2 \sum_{j<k} ln|z_j - z_k| + \frac{m}{2l_B^2} \sum_j |z_j|^2. \qquad (6.6.3)$$

β is identified as $1/m$. So Eq. (6.6.2) yields the density of the plasma. It should be remembered that it is only an analogy where a classical plasma constitutes of particles with charge m in a uniform (neutral) background. The existence of a plasma-like state at a very low temperature is counterintuitive, and is suggestive of a liquid phase of fermions. It would have been a crystalline state (for example, a Wigner crystal) at large values of the charge 'm'. It is indeed a new state of matter and often denoted as the Laughlin state.

In order to understand the potential V_{plasma}, it can be noted that the electric field, $\mathbf{E}(\mathbf{r})$, and the potential, $\phi(\mathbf{r})$, due to a point charge q are given by

$$\mathbf{E}(\mathbf{r}) = \frac{q\mathbf{r}}{r^2}, \text{ and } \phi(\mathbf{r}) = -qln(r). \qquad (6.6.4)$$

These yield Laplace's equation in 2D as

$$\nabla \cdot \mathbf{E} = -\nabla^2 \phi(\mathbf{r}) = 2\pi q \delta^2(\mathbf{r}), \qquad (6.6.5)$$

where $\delta^2(\mathbf{r})$ is the two-dimensional (2D) Dirac delta function. The first term in Eq. (6.6.3) is explained by the logarithmic dependence of the potential, which yields

$$V_{\text{plasma}}^{(1)} = m^2 \sum_{j<k} (-ln|z_j - z_k|). \qquad (6.6.6)$$

We are missing a factor of '2' here, but that can be absorbed in the definition of β (β can be redefined as $2/m$). The second term can be understood via noting that

$$\nabla^2 \frac{|z|^2}{4} = \frac{1}{l_B^2}, \text{ or, } \rho = -\frac{1}{2\pi l_B^2}, \tag{6.6.7}$$

where the charge density ρ satisfies Poisson's equation with a potential

$$V^{(2)}\text{plasma} = \frac{|z|^2}{4}. \tag{6.6.8}$$

This term denotes the energy of 'm' charges interacting with the negative charge density. It is obvious that $2\pi l_B^2$ is the area that contains one quantum of flux ($\Phi_0 = h/e$), which makes the background charge density to be B/Φ_0, which also denotes the density of flux in unit of the flux quantum.

While the Laughlin wavefunction demonstrates a lot of merit, one pertinent question still remains. How good is the wavefunction in practical cases? To enumerate its success, the overlap between the Laughlin state and the exact ones (obtained via exact diagonalization) for three particles and a few representative (odd) values of m and corresponding to a few different forms of the potential $V(\mathbf{r})$ (for example, $V(\mathbf{r}) = 1/r, -lnr, e^{-r^2/2}$) are obtained. The overlap between the two results is very close to 100%, thus ensuring that the proposed wavefunction is indeed a good one [93].

6.7 Composite fermions

To understand the concept of CFs, let us review the concept of vortices. In a reasonably different sense than the supercomputers (where magnetic flux lines puncturing a superconductor creates a vortex in a type-II material where the gradient in the phase of a condensate wave function produces circulating current that is detectable in experiments), vortices in FQHE can be understood as follows. A complex number $z(= re^{i\theta})$ has a vortex at the origin, implying that a complete circle around the origin creates a phase change of 2π. Similarly, the FQHE wave function contains $(z - z_0)^{2p}$, which implies particle '1' sees $2p$ vortices to be carried by particle '2' and vice versa. Thus, every particle can be imagined to carry an even number of vortices. The particles carrying the vortices are certainly not real. There are no real magnetic flux quanta attached to them. However, the intuitive picture leads us to the concept of CFs [92] and their relevance to the fractional statistics [98, 99].

A CF is defined as the bound state of a fermion and an even number of vortices or equivalently flux quanta (remember the equivalence of flux quantum and vortices come from the fundamental fact they are topologically similar in the sense they both result in a phase change of 2π upon circling about a closed path around it). The picture portrayed above is only for the convenience of visualization of a physical scenario; however, in reality there are no bound state of fermions and vortices (see Fig. 6.1).

Thus, a 2D system in the presence of a strong transverse magnetic field yields a picture where the electrons absorb a significant fraction of the external field and transform into the CF. Several experiments, such as thermopower [56, 57], Shubnikov-de Haas (SdH) oscillations and their cyclotron orbits [58], observation of Fermi sea [59] support the proposed CF picture. Importantly, the formalism of CF was capable of explaining the physics associated with the quantum Hall plateau observed at fractional filling of the LLs.

The role of interaction in the context of FQHE has been illustrated earlier, and interestingly, it is the only energy scale left in the problem (the kinetic energy of the electrons becomes an irrelevant constant). These strongly interacting electrons in the presence of a magnetic field, B, transform into such weakly interacting CFs in a much weaker field, B^*, where B^* is reduced from B (being absorbed by the electrons) by $2p\phi_0\rho$, namely[3]

$$B^* = B - 2p\phi_0\rho \qquad \begin{cases} 2p & : \text{an integer} \\ \phi_0 & : \text{flux quantum} \\ \rho & : \text{density} \end{cases} \qquad (6.7.1)$$

The situation is analogous to a filling fraction ν^* for the CF that corresponds to a fraction ν of the original electrons, and they are related to each other by

$$1 = \frac{\nu^*}{\nu} - 2p\nu^*$$

$$1 = \nu^*(\frac{1}{\nu} - 2p) \qquad (6.7.2)$$

$$\nu = \nu^*(1 - 2p\nu). \qquad (6.7.3)$$

---

[3]To arrive at the expression, consider the AB phase of a particle executing a circular motion of area A (disregarding the motion of all other particles) in an effective field B^* which is a result of $2\pi AB^*/\phi_0 \equiv 2\pi AB/\phi_0 - 2\pi(2p)\rho A$, where the last term denotes the flux due to $2p$ vortices.

Equivalently, one can write

$$\frac{1}{\nu^*} = \pm(\frac{1}{\nu} - 2p).$$ (6.7.4)

The CF picture not only preserves the fundamental property of fermions, such as the exchange of fermions involving a change of sign, the Aharonov–Bohm phase associated with the cyclic variation of the wave function around a close path yields a value 'one'.[4]

However, there is an important development that has occurred here. The highly degenerate many-particle ground state for fractional filling $(\nu < 1)$ in the absence of interaction transforms into the ground state of the CF with drastically reduced degeneracy corresponding to a filling $\nu^* > 1$. For integral values of ν^*, the situation yields a non-degenerate ground state. The loss of degeneracy makes the interaction among the CF to be vanishingly small. Thus, the FQHE of the original electrons transforms into IQHE of the CF and should yield a great simplification to a rather complex problem. The wavefunction for the CF can be obtained from the same variational state written down by Laughlin as in the following,

$$\psi_\nu = \prod_{j<k}(z_j - z_k)^{2p}\phi_{\nu^*},$$ (6.7.5)

where ϕ_{ν^*} denotes wavefunction for the non-interacting electrons a filling ν^* and $(z_j - z_k)^{2p}$ is the similar Jastrow factor that was present in Laughlin's proposal, which ensures keeping the CF apart, and no two of them come close to one another.[5]

The success of the CF picture can be tested by plotting the magneto (longitudinal) resistivities as a function of the inverse filling fraction $\frac{1}{\nu^*}$, which is proportional to B^*. A comparison between the IQHE and FQHE resistivities demonstrates the dips in the resistivity (see Figs. 1.3 and 1.4) correspond to integer filling fractions, while the ones in the lower panel correspond to the fractional fillings but are related by

$$\nu = \frac{\nu^*}{2\nu^* + 1},$$ (6.7.6)

which can be obtained by putting $p = 1$ in Eq. (6.7.4). Thus, the IQHE of CF occurs at filling fractions $\nu^* = n$, where n is an integer. The filling fractions, despite being

---

[4]A phase $e^{2\pi\phi/\phi_0}$ with $\phi = 2p\phi_0$ yields, $e^{2\pi(2p)} = 1$.

[5]The CF function vanishes as $\nu^{2(2p+1)}$ instead of ν^2 that is familiar for electrons obeying Pauli principle.

very different, look quite similar. This similarity re-emphasizes the dynamics of the interacting electrons resembles that of the non-interacting CF very closely at a reduced magnetic field B^*.

Many of the observed fractional values for the quantum Hall plateaus can be obtained by putting $p = 1, 2, 3 \ldots$ in Eq. (6.7.4), which can be viewed potentially as in Fig. 6.1.

Other fractions can be obtained by putting a negative sign in the following equation,

$$\frac{1}{\nu^*} = \pm\left(\frac{1}{\nu} - 2p\right), \tag{6.7.7}$$

where the negative sign above implies an effective field B^* that is directed antiparallel to the applied field B. A few observed fractions and the corresponding (even) number of flux quanta (vortices) are tabulated in Table 6.1. Direct numerical solution of the Schrödinger equation for 10–15 particles in the presence of a pairwise Coulomb potential of the form

$$\mathcal{H} = \frac{1}{4\pi\epsilon_0} \sum_{j \neq k} \frac{e^2}{|\mathbf{r}_j - \mathbf{r}_k|} \tag{6.7.8}$$

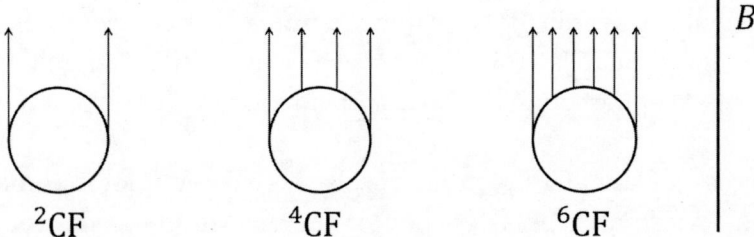

FIGURE 6.1. Electrons capture 2, 4 and 6 flux quanta, and the composite particles are known as composite fermions (CFs).

TABLE 6.1. A few representative p values and the corresponding ν values are shown.

p	$\nu = \frac{\nu^*}{2p\nu^* + 1}$
1	$\frac{1}{3}, \frac{2}{5}, \frac{3}{7} \cdots$
-1	$\frac{2}{3}, \frac{3}{5}, \frac{5}{7} \cdots$
2	$\frac{1}{5}, \frac{2}{9} \cdots$
-2	$\frac{2}{7}, \frac{3}{11} \cdots$

has been solved exactly.[6] Corresponding to ν^* equal to an integer (say, n) one expects a gap in the spectrum. At such values of filling, the Coulomb interaction lowers the enormous degeneracy of the LLL by a great extent, so as to produce a non-degenerate ground state. This state notionally denotes $\nu^*(= n)$ filled LLs of CF. For the electrons, the filling fraction is of course ν, which denotes a fraction (with odd denominator). Further, the wavefunctions of this CF have an excellent overlap with the wavefunctions stated in Eq. (6.3.3).[7] In order to recover Laughlin's wavefunction, consider $\nu^* = 1$, which denotes $\nu = \frac{1}{2p+1}$. The wavefunction corresponding to the filling ν is

$$
\begin{aligned}
\psi_{\nu=\frac{1}{2p+1}} &= \prod_{j<k}(z_j - z_k)^{2p}\Phi_{\nu^*=1} \\
&= \prod_{j<k}(z_j - z_k)^{2p}\left[\prod_{j<k}(z_j - z_k)\, e^{-\frac{1}{4l_B^2}\sum_l |z_l|^2}\right] \\
&= \prod_{j<k}(z_j - z_k)^{2p+1}\, e^{-\frac{1}{4l_B^2}\sum_l |z_l|^2}.
\end{aligned}
\tag{6.7.9}
$$

This is exactly Laughlin's wavefunction if one identifies $m = 2p + 1$.

Thus far, we have identified that the challenges of FQHE are too many, and most importantly it has to do with the absence of any 'small' parameter of the problem. The carriers are frozen, and hence there is no kinetic energy, leaving the interparticle interaction to be the only energy scale of the problem. This impedes known methods of solution to be applicable. Further, the enormous degeneracy of the LLs (without invoking the Coulomb interaction) aggravates the problem. A third, and quite a crucial one, is the unavailability of a so-called *normal state*, that is, there is no known state that becomes unstable in favour of a fractional quantum Hall state by turning on a weak interparticle interaction. Hence, there is no small expansion parameter; humongously large number of ground states that are degenerate in the absence of interaction, and the non-existence of a normal fluid state put together narrate about an enormously complicated problem.

However, this enormously complicated problem is intuitively solved by writing wavefunction (all credits to Laughlin), which we have denoted by $|\psi(\nu)\rangle$. Using

---

[6]Since the Coulomb potential is the only energy scale left in the problem, we have represented it by the Hamiltonian, $\mathcal{H}$.

[7]For an excellent discussion on the subject, see Ref. [92].

this wavefunction, the energy eigenvalues for the problem can be written as

$$\langle \mathcal{H} \rangle = E_v = \langle \psi_v | \sum_{j<k} \frac{1}{|r_j - r_k|} | \psi_v \rangle + V_{e-b} + V_{b-b}, \tag{6.7.10}$$

where the first term on the RHS of Eq. (6.7.10) is the Coulomb energy, and the last two denote the interaction energy of the electron-background and between the background entities, such as ions. Importantly for us, the LLL index v is given by

$$v = \frac{v^*}{2pv^* \pm 1}, \tag{6.7.11}$$

where the index v^* is defined for the wavefunctions of the non-interacting electrons, namely ψ_{v^*}. Thus, a strongly interacting problem in the presence of an external magnetic field B is reduced to a non-interacting one in an effective field given by

$$B^* = B - 2pv\Phi_0. \tag{6.7.12}$$

Thus, B^* can be zero if B is exactly cancelled by the second term in Eq. (6.7.12), thereby leading to a non-degenerate state (since the degeneracy is proportional to B), or even be negative, where the vortices carried by composite particles point opposite to the applied field. Further, ψ_v contains a Jastrow factor, namely $\Pi_{j<k}(z_j - z_k)^{2p}$, which projects the wavefunction onto the LLL.

    Let us consider the following example. A correspondence between the integer quantum Hall ground state with $v^* = 3$, that is, three filled levels and the corresponding fermion picture with $v = \frac{v^*}{2pv^*+1}$ is shown schematically in Fig. 6.2.

    The above discussion yields an intuitive picture of CFs that are *bound states* of fermions and $2p$ 'vortices' (p is an integer). By absorbing the 'vortices' or the flux quanta, the electrons minimize the interparticle interaction energies. As these

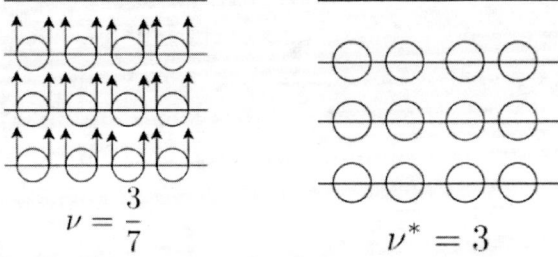

FIGURE 6.2.   The schematic plots show the integer quantum Hall effect for $v^* = 3$ (right) and the fractional quantum Hall effect for $v = \frac{3}{7}$ (left).

composite structures (particles + vortices) propagate in a quantum Hall fluid, they create irreducible phases, which (partially or fully) cancel the phase due to the external magnetic field.

The CFs truly are new entities, and not seen before. Since a vortex has a quantum mechanical origin, the CFs are quantum objects (owing to the quantum mechanical phases associated with the vortices) as well. However, in a fluid they behave as free particles. Cooper pairs, which are bound states of two electrons formed in the presence of lattice excitations, that is phonons, bring up a close analogy, but are distinct in many ways as well. In fact, the topological quantization of the vorticity is directly linked with the quantization of the Hall plateaus.

We prefer to stop here on the topic of CFs, and suggest more specialized articles and books on the subject. Particularly, the book by J. K. Jain [92] has given an extensive account of CFs and in general, on the theory of FQHE in a fairly lucid manner.

6.8 Hierarchy approach to FQHE

We have so far discussed the Laughlin wavefunction and a CF scenario to understand the physics of the fractional quantum Hall fluid. There are significant merits of both these approaches. The Laughlin wavefunction is similar to the Jastrow function earlier employed to the superfluids, such as ^{4}He. A number of fractions that are experimentally observed could be explained by the Laughlin states, while a very large number of them remained elusive. On the other hand, the CF ideas were motivated by considering FQHE to be analogous to IQHE, which eliminates the effects of strong interaction among the fermions by absorbing $2p$ flux quanta. Several experiments corroborate the CF picture.

The non-Laughlin fractions still needed an explanation. Besides, we have a ground state wavefunction in some form, and thus a natural expectation is to obtain information on the excited states. This forms the foundation of the hierarchy scenario, which assumes Laughlin wavefunction as the starting point. The construction of a hierarchical wavefunction was put forward by Haldane [94] and Halperin [95] to provide justification to the non-Laughlin fractions, that is, the ones that are not of the form $1/m$. The idea is to create '*Laughlin-like daughter states*' from a given '*parent state*'. Additional states with new fractions are iteratively generated from the Laughlin fractions. Thus, all the odd-denominator fractions are reproduced. The fractions at any level of the hierarchical scheme are represented as continued fractions.

A physical picture relevant to the hierarchy description may be enunciated as follows. At the centres of the quantum Hall plateau, which corresponds to an incompressible fluid with uniform density. As the magnetic field is ramped up, localized quasiparticles or quasiholes develop corresponding to excess or deficit of densities that are initially pinned to the impurity centres. However, as the external field becomes larger, the excitations split from the impurities and become mobile, leading to a loss of the quantized Hall response. These quasiparticles (or the quasiholes) being charged, they may be considered to condense into Laughlin-like states on top of the original quantum Hall fluid. The scenario can be iterated which will lead to arbitrary filling fractions (basically the non-Laughlin fractions) with odd denominators.

To understand how the continued fractions arise, it is useful to review the excited states of the quantum Hall fluid, that is, the quasiparticles and the quasiholes. The quasiparticles denote excitations, each of which carries a charge $-\frac{e}{m}$, while the quasiholes denote those with charge $\frac{e}{m}$. These excitations behave as individual particles, which are non-interacting, each with a fractional charge; however, the total charge adds up to an integer, as it should be.

Halperin [96] modified the Laughlin wavefunction and made an ansatz for including the quasiparticle and the quasihole excitations as

$$\psi = P(z_k)Q(z_k)e^{-\sum_j |z_j|^2/4l_B^2},\qquad(6.8.1)$$

where $Q(z_k)$ accounts for the quasiparticle and the quasihole excitations and is given by

$$Q(z_k) = \Pi_{j<k}^N(z_j - z_k)^{\pm 1/m}\qquad(6.8.2)$$

and $P(z_k)$ is the ubiquitous Jastrow factor given by

$$P(z_k) = \Pi_{j<k}^N(z_j - z_k)^{2p}.\qquad(6.8.3)$$

Thus,

$$\psi = \Pi_{j<k}^N(z_j - z_k)^{2p\pm 1/m}e^{-\sum_j |z_j|^2/4l_B^2}\qquad(6.8.4)$$

with m still being an odd integer.

Put together $P(z_k)$ and $Q(z_k)$ give rise to an interchange that is distinct than that of the antisymmetric exchange of fermions. It is further evident that the maximum angular momentum of the state is

$$|J_{max}| = N\left(2p \pm \frac{1}{m}\right).$$

The filling fraction can be obtained by noting that the area A to be given by

$$A = N\left(2p \pm \frac{1}{m}\right)(2\pi m l_B^2).$$

This yields the number of states within an area A, which is given by

$$\mathcal{N} = \frac{BA}{\Phi_0} = \frac{\Phi}{\Phi_0} = \left(2p \pm \frac{1}{m}\right)m^2 N.$$

Hence, the filling fraction is

$$\nu = \frac{1}{2pm^2 \pm m}. \tag{6.8.5}$$

The above expression allows all fractions and not only the Laughlin $(1/m)$ types. At the third level of the hierarchy, the filling fractions are denoted by

$$\nu = \frac{1}{m \pm \cfrac{1}{2p_1 \pm \cfrac{1}{2p_2 \pm \cdots}}}, \tag{6.8.6}$$

For example, for $p_j = 1$ and $m = 3$, at the third level one gets

$$\nu = \frac{1}{3 \pm \cfrac{1}{2 \pm \frac{1}{2 \pm \cdots}}}. \tag{6.8.7}$$

For the $+$-branch, one gets $3/11$ and $5/17$, and the $-$-branch, the fractions are $3/7$ and $5/13$. Schematically, the hierarchy at this level is expressed via Fig. 6.3.

6.9 Fractional statistics

The occurrence of quantization of the Hall plateaus at fractional filling in the expression bears a testimony for the quantization of the fractional charge. Ironically, the fractionally charged quasiparticles being localized at the quantum Hall plateaus do not conduct and the contribution to the Hall current is due to the background (neutral) state which is incompressible and does not contain any quasiparticles. Nevertheless, the fractional charge requires a particle statistics just

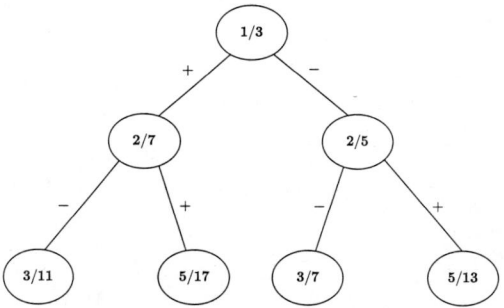

FIGURE 6.3.   Schematic plot showing the hierarchy scheme for $m = 3$ and $p_j = 1$. At the third level, one obtains non-Laughlin fractions, such as 3/11, 5/17, 3/7 and 5/13.

as the integral (including zero) charge particles. It should be kept in mind that the particle statistics shows up in the form of collective phenomena, such as formation of a Bose–Einstein condensate, or a degenerate Fermi gas.

In this context, we may recall that we have been exposed to the concept of identical particles, which are either Bosons or Fermions and obey the properties of symmetry or antisymmetry respectively under the exchange of particles pairwise. That is, for two particles,

$$\psi(\mathbf{r}_1, \mathbf{r}_2) = \pm \psi(\mathbf{r}_2, \mathbf{r}_1), \tag{6.9.1}$$

where the + sign refers to Bosons and − sign is applicable to Fermions. However, it is also true that probabilities are the same, namely

$$|\psi(\mathbf{r}_1, \mathbf{r}_2)|^2 = |\psi(\mathbf{r}_2, \mathbf{r}_1)|^2. \tag{6.9.2}$$

Such that the wave functions, upon exchange of particles, at most picks up a phase,

$$\psi(\mathbf{r}_1, \mathbf{r}_2) = e^{i\pi\alpha}\psi(\mathbf{r}_2, \mathbf{r}_1). \tag{6.9.3}$$

Repeating the exchange brings back the same state, which implies $e^{2\pi i\alpha} = 1$, where $\alpha = 0$ for Bosons and $\alpha = 1$ for Fermions. There is a subtle point here which is not explicitly stated. In three dimensions, a rotation by 2π brings us back to the original state which should be equivalent to changing a pair of particles twice. Such exchanges (or equivalently a rotation by 2π) are continuously, also called world lines, connected and do not cross each other's path. However, in two dimensions, there is a big difference where such paths cross and wind around each other. For example, there is a distinction between a clockwise and an anticlockwise exchange of particles, though in either case, their paths get tangled and form braids. A braid describes a pattern formed by the interaction of two or more strands of wire (or hair). Thus, a clockwise braid is distinct from an anticlockwise braid and is said to belong to different topological sectors.

A tangled path implies an arbitrary phase involved in the exchange of particles because the paths cross, there arises an ambiguity in the phase and hence α can assume any value in the interval $[0:1]$. More concretely, say, for an anticlockwise exchange,

$$\psi(\mathbf{r}_1, \mathbf{r}_2) = e^{i\pi\alpha}\psi(\mathbf{r}_2, \mathbf{r}_1). \tag{6.9.4}$$

A clockwise exchange results in

$$\psi(\mathbf{r}_1, \mathbf{r}_2) = e^{-i\pi\alpha}\psi(\mathbf{r}_2, \mathbf{r}_1). \tag{6.9.5}$$

These particles are known as anyons owing to their allegiance to 'any' statistics that interpolates between bosons and fermions. 'Any' statistics here refers to a fractional statistics and applies to the quasiparticles (or more aptly called the quasiholes). The charge of the quasiparticles or the anyons is fractional.

Without going into elaborate calculations, we refer to published works by D. Arovas, J. Schrieffer and F. Wilczek [99] and D. Tong in 'The Quantum Hall Effect' [69]. The Berry phase around a closed loop for that quasiparticles can be obtained as

$$\Phi_B = \frac{e\Phi}{mh} = \frac{\Phi}{m\Phi_0} \qquad (\text{m : odd}), \tag{6.9.6}$$

where Φ is the flux enclosed by a quasiparticle around a closed contour and $\Phi_0 = $ the flux quantum (h/e). It also has the interpretation that

$$\Phi_B = \frac{e^*\Phi}{h}, \tag{6.9.7}$$

where e^* refers to a fractional charge.

6.10 Non-abelian anyons

Whenever we talk about particle statistics, the concept of parity of the wavefunction under the interchange of particles is invoked. As quantum particles are identical, no physical observable should change under the permutation of the particle indices. However, it does not require the wavefunction to be invariant under such a permutation, as the physical observables depend on the absolute value (mod square) of the wavefunction, and not on phase. The fermions pick up a negative sign when a pair of particles are exchanged, while the bosons pick up a '+'sign, or in other words, they pick up no sign at all. Thus, two such interchanges of two pairs of particles do not pick up any sign at all in either of the cases. Equivalently stated, if one particle is taken around another one in a closed path, the scenario is topologically equivalent to a process in which neither of the particles moves at all. The wavefunction changes by a '+'sign (for bosons) and '−'sign (for fermions) for one such move, or no sign at all for two such moves around each other.

However, there seem to be no physical grounds on why (and how) such symmetric (+sign) and antisymmetric (−sign) wavefunctions originate. The ideas of particle exchange or permutation of the particle indices appear quite artificial, unless a physical meaning is ascribed to it. Leinaas and Myrheim [98] have associated a physical meaning to these permutations, where an exchange is contemplated as an adiabatic and continuous transformation where the particle exchange happens along continuous paths. The above situation is usually narrated in 3D, albeit not stated explicitly anywhere. In 3D, two particles can encircle each other, without crossing each other's path, which is quite different in 2D. The path followed by one particle in encircling the other cannot be done without crossing the trajectories of each other. When two particles are interchanged twice, say in a clockwise manner, the paths along which such a process can be done yield a non-trivial winding. As a consequence, quite unlike 3D, the final state may not be the same as the initial one. Thus, the quantum systems in 2D are qualitatively different than those in 3D. In fact, the quantum statistics in 1D is not well defined as the process of interchange of particles is impossible without going through each other.

Leaving aside any discussion in 1D, let us rephrase the earlier discussions in 2D and 3D. The essential difference between particle exchanges in 2D and 3D may be realized for a simple system comprising two particles. In 3D, let us place the origin of the coordinate system to be at the location of one of the particles. Now an exchange would imply that a second particle traverses around the origin. Exchanging twice implies forming a loop around the origin. In 3D, the space is simply connected, and the loop may be continuously deformed to its original configuration with the particles never coinciding. Hence, after exchanging twice, the wavefunction should re-acquire its original value. In 2D, this is not possible, that is, no number of exchanges generate a loop deformable to a point. Further, unlike in 3D, the particles pick up any arbitrary phase between 0 and π. Because of this *'any'* phase (and not just 0 and π) earned them the name *'anyons'* (first called by Wilczek [99]).

Mathematically, a two particle exchange, say in a counterclockwise manner of the anyons is denoted by

$$\psi(\mathbf{r_1}, \mathbf{r_2}) = e^{i\theta} \psi(\mathbf{r_2}, \mathbf{r_1}). \tag{6.10.1}$$

Two such counterclockwise exchanges result in

$$\psi(\mathbf{r_1}, \mathbf{r_2}) = e^{2i\theta} \psi(\mathbf{r_1}, \mathbf{r_2}), \tag{6.10.2}$$

where $\theta \neq 0$, or π. Moreover, there can be two types of anyons, namely the abelian and non-abelian anyons. In the abelian variety, the order of exchange of particles leaves the corresponding wavefunction unchanged, while for the non-abelian anyons, the order of exchange matters, and the wavefunction picks up a sign.

From the context of QHE it is evident that indistinguishable particles in 2D are important for our discussion. The discussion of IQHE did not necessitate deliberation on the statistics discussed above, since it can be explained on the basis of single particle picture. However, FQHE indeed involves Coulomb interaction, and hence the description requires invocation of a many particle state which forces the particle statistics to be important. Thus, anyons will have a physical realization in a system demonstrating FQHE. Furthermore, the non-abelian anyons are serious candidates for fault tolerant quantum computation, or the so-called *topological quantum computation* [100].

The mathematical formalism of the statistics for particles in 2D requires taking into account of the features elaborated above. It thus involves a fundamental group, which owing to the adiabatic nature of exchange of particles is unitary. The group is called the *'braid group'*. The discussion on anyons is directly linked to the representation of this braid group.

6.11 The braid group

Now consider an N-particle system which from an initial configuration $\{r_1, r_2.....r_N\}$ goes to a final configuration $\{r'_1, r'_2.....r'_N\}$, where the primed coordinates comprise of permutations of all or some of the particles. Each of the permutations can either be clockwise or anticlockwise and consist of elements of the braid group, $\mathcal{B}_N$. $\mathcal{B}_N$ is a symmetric group and an infinite one. To understand the properties of the group, let us define the anticlockwise and clockwise permutations of the ith and the $(i+1)^{th}$ particle by σ_i and σ_i^{-1} for

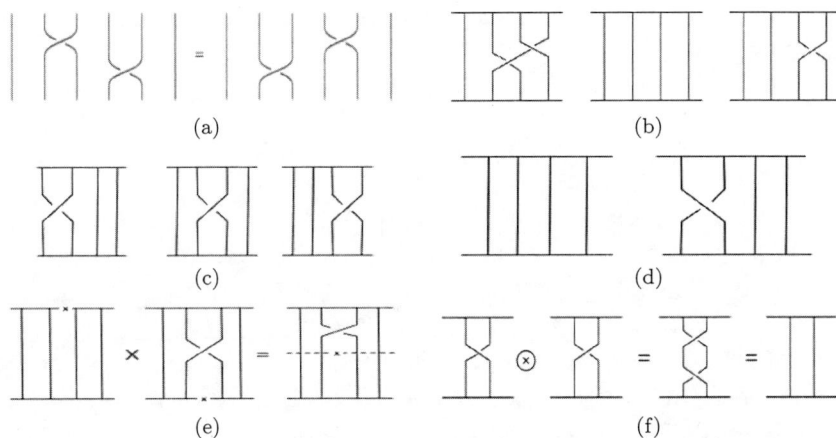

FIGURE 6.4. (a) The figure corresponds to the braid diagram showing $\sigma_i\sigma_{i+2} = \sigma_{i+2}\sigma_i$. (b) and (c) respectively correspond to the elements and the three generators of the braid group $\mathbb{B}_4$. (d) The figure corresponds to the identity and the inverse of the generator σ_1. (e) Group multiplication method is shown schematically. (f) Here, the product of σ_1 and $(\sigma_1)^{-1}$ is shown. It gives rise to identity.

$1 \leq i \leq N^8 - 1$. The exchange of two disconnected pairs of particles should be independent of one another, and thus the order of these exchanges should not make a difference. In mathematical notations, this is equivalent to

$$\sigma_i\sigma_j = \sigma_j\sigma_i \text{ for } |i - j| \geq 2. \tag{6.11.1}$$

A valid commutator structure for these elements requires a Yang–Baxter constraint [101], namely

$$\sigma_i\sigma_{i+1}\sigma_i = \sigma_{i+1}\sigma_i\sigma_{i+1}. \tag{6.11.2}$$

In general, for the anyons $\sigma_i \neq 1$. A visual aid of the group properties given in Fig. 6.4 may help. Next, we discuss abelian and non-abelian anyons.

The abelian anyons correspond to the 1D representation of the braid group, $\mathcal{B}_N$. The representation is denoted by

$$\pi(\sigma_k) = e^{i\theta}. \tag{6.11.3}$$

Because of the above structure $\pi(\sigma_k)$ mutually commute, which implies that the resulting statistics is abelian. Let us denote the arbitrary permutation of the particle

---

[8]A counterclockwise exchange indeed annihilates a clockwise exchange, and hence one is the inverse of the other.

indices, namely

$$(\mathbf{r}_1, \mathbf{r}_2....\mathbf{r}_N) \implies (\mathbf{r}'_1, \mathbf{r}'_2....\mathbf{r}_N).$$

Let this be an element α of the abelian braid group, $\mathcal{B}_N$, such that

$$\pi(\alpha) = e^{i(m_1 + m_2 + ...m_N)\alpha}. \tag{6.11.4}$$

Hence, one gets the permuted wavefunction, $\psi(\mathbf{r}'_1, \mathbf{r}'_2....\mathbf{r}'_N)$, from the original wavefunction, $\psi(\mathbf{r}_1, \mathbf{r}_2....\mathbf{r}_N)$, by acting $\pi(\alpha)$ on the latter. That is,

$$\psi(\mathbf{r}'_1, \mathbf{r}'_2....\mathbf{r}'_N) = \pi(\alpha)\psi(\mathbf{r}_1, \mathbf{r}_2....\mathbf{r}_N) \tag{6.11.5}$$

$$= e^{i(m_1 + m_2 + ...m_N)\alpha}\psi(\mathbf{r}_1, \mathbf{r}_2....\mathbf{r}_N).$$

The above equation implies that $m_1 + m_2 + ...m_N$ is the effective permutation that takes $\psi(\mathbf{r})$ to $\psi(\mathbf{r}')$ (particle indices are suppressed). Further, each of the m_i denotes a counterclockwise exchange of i and $i + 1$ indices minus the number of clockwise exchanges of these indices required to go from an initial to a final state. Permuting any of the m_i indices does not change the statistics.

However, the operator representation of the braid group in higher dimensions is in general not an abelian one, that is, the elements do not commute. Let $\tilde{\pi}(\sigma_k)$[9] denotes some representation of $\mathcal{B}_N$. $\tilde{\pi}(\sigma_k)$ is an $m \times m$ square matrix, denoted as the braiding matrix, which enters through the transformation equation of the basis states under the exchange via

$$\psi_i = \sum_j [\tilde{\pi}(\sigma_k)]_{ij}\psi_j, \tag{6.11.6}$$

where $[\tilde{\pi}(\sigma_k)]_{ij}$ denotes the ij^{th} element of the braiding matrix. The above equation has an important ramification which states that an exchange truly creates a non-trivial transformation in the many-particle Hilbert space. The statistics applies to excitations of the systems that demonstrate FQHE, particularly for certain pathological fractions, especially the ones with even denominator. Obviously, an even denominator aims to threaten the basic premise of the Laughlin states (with odd denominators due to fermion statistics) that the discussion of FQHE is based on.

The above discussions on anyons and 2D statistics (also called θ-statistics), albeit brief allows us to connect smoothly to the ongoing discussion of FQHE

---

[9] $\tilde{\pi}$ is used to distinguish it from the representation π used for the abelian case.

through the realization of non-abelian anyons that aids us in understanding the even denominator fractions (such as 5/2) realized experimentally.

A little elaboration of this anomalous fraction would benefit the readers. Around the year 1987, GaAs/AlGaAs heterostructure samples started showing even larger mobility ($\sim 10^6$ cm^2/Vs) than before. In the same year, Willet *et al.* [102] reported the first evidence of an even denominator ($\nu = 5/2$) plateau in the Hall resistivity.

The FQHE state at $\nu = 5/2$ contradicts the other plateaus obtained of the form $\nu = p/(2p + 1)$, and is strongly believed to denote a non-abelian quantum Hall state. At this filling fraction, the LLL is completely filled with both spin-↑ and spin-↓ electrons, and the second Landau level (SLL, also referred to as the first excited LL) is at $\nu = 1/2$. In the CF language, the latter means that there are exactly two flux quanta per electron. Ignoring what happens in the LLL, one may focus only on the SLL. The CFs in SLL may be assumed to renormalize the inter-particle interactions. Numerical results suggest that the CFs in the SLL form bound pairs (such as the Cooper pairs), and the resulting state Bose condense into a superfluid-like state. Thus, for the $\nu = 5/2$ state, the excitations are the vortices, analogous to what happens in the mixed state of a superconductor. Each of the vortices is a zero energy Majorana mode and carries a fractional charge $e/4$ and has the properties of an anyon. These anyons are quasiparticle excitations in the bulk of a $\nu = 5/2$ FQHE state. The low-energy description of such a state requires the knowledge of conformal field theory which is beyond the ambit of this book.

We have provided a brief description of the particle statistics in 2D, and underscored its distinct nature than that in 3D. This has led us to the discussion on anyons and their representation, namely the braid group. Having stated the preliminary properties of the braid group, we have discussed the relevance of anyons, and in particular, of the non-abelian anyons to the $\nu = 5/2$ FQHE state.

6.12 Fractional quantum Hall effect in graphene

This section should come with the disclaimer that the subsequent discussion is only meant to touch upon the early experimental discovery of FQHE in graphene, and by no means provides a complete description of the topic. The interested readers are encouraged to look at more specialized articles and reviews on the subject.

It is fairly well established that the π electrons in graphene have a large kinetic energy, which is why the correlation effects can safely be ignored and a tight-binding description works. A quick recap of the Hall effect in graphene (which we have elaborately discussed in Chapter 4) will be helpful to the readers. In the presence of a strong magnetic field, IQHE in graphene demonstrated unusual filling fractions (with respect to the 2DEG) given by $\nu = 4(n + \frac{1}{2})$, n being an integer including zero. This gives rise to plateaus at $\nu = \pm2, \pm6, \pm10.....$ The sequence of plateaus originates from two distinct features, namely the two-fold spin and two fold valley (together making it four fold) degeneracies of the zeroth LL.

Shortly afterwards, in 2009 two experimental groups from Rutgers (USA) [103] and Columbia (USA) [104] observed FQHE in suspended graphene. Afterwards, a number of groups observed the same effect with different fractions. For example, see Ref. [105]. It was also observed in superlattices, such as in heterostructures of h-BN/graphene/h-BN [106].

In both the early discoveries (in 2009) on clean and suspended graphene, a plateau at $\nu = \frac{1}{3}$ is observed, a fraction which is also observed in the case of 2DEG, now with the pseudospin in graphene playing the role of actual spin in 2DEG. In addition, other fractions, namely $\nu = \frac{2}{3}, \frac{2}{5}, \frac{3}{5}, \frac{3}{7}, \frac{4}{7}, \frac{4}{9}$ are also observed via the measurement of inverse compressibility[10] as a function of the carrier density and the magnetic field [113]. These fractions are also present in 2DEG. Further, the study of temperature effects yields interesting results. For example, the $\nu = \frac{1}{3}$ plateau disappears beyond a few Kelvin, suggesting that the electron–electron interaction loses prominence due to thermal effects.

The origin of the fractional plateaus is attributed to the enhanced electronic interaction in a suspended graphene owing to poor dielectric screening. The dielectric constant, ϵ, makes an entry into the interaction term via, $V_{e-e} = \frac{e^2}{\epsilon l_B}$ ($l_B = \sqrt{\frac{\hbar}{eB}}$). ϵ is of the order of 1 in suspended graphene, compared to $\epsilon \simeq 13$ in GaAs, GaAlAs heterostructures. Clearly, the electron–electron interaction is large in graphene.

Other than the fractionally quantized plateaus, integer plateaus are observed at $\nu = 0, \pm1$ [107, 108] in ultra-clean samples (so that disorder effects are subdued) where quantum limits can be attained at relatively moderate values of the external field and are unusual enough with regard to the formula for the filling fraction quoted earlier. These quantum Hall states are fragile and are thought to be arising out of lifting of the spin and valley degeneracies induced by electron–electron

---

[10]Compressibility (κ) is defined as $\kappa = \frac{dn}{d\mu}$, where n denotes the carrier density and μ is the chemical potential.

interaction effects. These states also yield divergent resistivity at very large magnetic fields.

It is also conjectured that the pathological integers (for example, $\nu = 0, \pm 1$) may have resulted in the lifting of the four fold degeneracies. The lifting of the spin and valley degeneracies is an intricate matter and can be done in two ways [103]. In the first case, the spin degeneracy is lifted first followed by the valley degeneracy. In the other, the reverse happens, that is, the valley degeneracy is lifted ahead of the spin degeneracy. The former results in a quantum Hall ferromagnet [109, 110], while the latter yields magnetic catalysis [111, 112]. While both cases result in gapped (insulating) bulk, only the former (where spin degeneracy is lifted first) supports topological character with counter propagating edge modes traversing the Fermi energy. The latter scenario supports trivial insulating states with no edge modes. Interestingly in a quantum Hall ferromagnet, the spin and the valley degeneracies can be lifted for all the LLs, and thus will result in all integer filling fractions being allowed.

As an end note, it can be said that the observation of FQHE in graphene requires ultraclean and suspended samples (the situation remains inconclusive if the above conditions are not strictly met) promotes investigation of strong electronic correlations and interplay of interaction and symmetries of fractionally quantized states. Besides, it opens up avenues to study LL mixing induced by disorder and interaction effects.

Epilogue

Quantum Hall states are the first examples of topological insulators that demonstrate completely contrasting electronic behavior between the bulk and the edges of the sample. The bulk of the system is insulating, while there exists conducting states at the edges. Moreover, the Hall conductivity is quantized in units of a universal constant, e^2/h. It became clear later on that the quantization is actually related to a topological invariant known as the Chern number. The geometric interpretation of this invariant is provided by the Gauss–Bonnet theorem, which relates the integral of the Gaussian curvature over a closed surface to a constant that simply counts the number of 'genus' (or holes) of the object. In solid state physics, the closed surface is the Brillouin zone, and the Gaussian curvature is analogous to a quantity known as the Berry curvature, integral of which over the Brillouin zone yields the quantization of the Hall conductivity.

In Chapter 1, we begin with a historical overview of the quantum Hall effect. The experiment and the physical systems are described with an emphasis on the two-dimensional (2D) nature of the 'dirty' electronic system in the presence of a strong perpendicular magnetic field at low temperature. The Hall resistivity as a function of the field shows quantized plateaus in unit of h/e^2 with an accuracy of one part in more than a billion. Very surprisingly, the longitudinal resistivity synergetically vanishes at the positions of the plateaus for the Hall resistivity. This indicates the emergence of a phase with an inherent ambiguousness of being a perfect conductor and a perfect insulator at the same time. However, such an ambiguity can only be reconciled for an electron gas confined in a plane in the presence of a magnetic field.

Quite intriguingly, the presence of the perpendicular magnetic field introduces 'another' quantization, which replaces the band structure (energy as a function of the wavevector) of the electronic system. This quantization was shown via solving the Schrödinger equation in the presence of a Landau gauge. The resultant energy levels of this problem are the infinitely degenerate Landau levels, which slightly broaden due to the presence of impurity and disorder but still remain distinct and cause quantization of the Hall conductivity as the magnetic field is ramped up gradually. Further, this quantization is visioned as a quantum pump by Laughlin where an electron gas in a planar disk geometry subjected to a magnetic field shows a transfer of one unit of charge from the inner to the outer edge of the disk as the magnetic flux changes by one quantum ($= \frac{h}{e}$).

In Chapter 2, we have digressed a little bit and went on to discuss discrete symmetries, such as inversion symmetry, time reversal symmetry, particle–hole symmetry and chiral symmetry, and examined how the Hamiltonian transforms under such symmetry operations. These symmetries yield protection to the topological phases.

In Chapter 3, we have discussed three paradigmatic models that show robust topological features that are resilient to local perturbations that do not violate the relevant symmetries. In this connection, we have studied two one-dimensional (1D) models, namely the Su–Schrieffer–Heeger (SSH) model and a Kitaev chain with superconducting correlations. Further, a quasi-1D ladder system, known as the Cruetz ladder, is discussed, which shows topological features. The topology in the SSH model is known to be induced by a dimerized hopping with two atoms per unit cell and is stabilized by the chiral symmetry of the Hamiltonian. As long as the chiral symmetry is intact, and when the intra-cell hopping amplitude is larger than the inter-cell one, the model displays localized zero modes at the edges. The topological phase is further characterized by a finite value of the winding number. A similar scenario is presented by the Kitaev model, which involves spinless fermions coupled by p-wave superconducting correlations on a tight-binding chain. The topological property of the model is protected by the particle–hole symmetry that is inherent to superconductors. Further, similar to the SSH model, the topological phase is characterized by zero modes at the edges of an open chain and a finite value for the winding number. Interestingly, the zero mode is twofold degenerate, which is inseparable and differs by fermion parity. Moreover, these states have a formal similarity with the Majorana fermions, which correspond to their own antiparticles. Hence, we discuss the Cruetz ladder, which has a rich parameter space defined by the hopping terms along the leg, rungs and diagonals, along with a magnetic flux threading it. In a certain parameter regime, the model shows topological properties with robust zero energy edge modes. A reason for us to study this rather rarely studied system is the following. It is neither a 1D model, nor 2D, and hence challenges the conventional periodic table of topological insulators based on the tenfold classification. Finally, for completeness, we have included a brief discussion on the classification of the topological materials according to the schemes introduced by Altland and Zirnbauer. This aids us in identifying the class of the topological systems discussed in this chapter and the corresponding topological invariants.

In Chapter 4, we discuss graphene, which is a prototype 2D material and often a hobby horse owing to the electrons demonstrating massless Dirac dispersion at low energies. We also study the quantum Hall effect in Graphene, which aids in distinguishing between the behavior of the relativistic and the non-relativistic (as in 2DEG) electrons in the presence of a strong magnetic field. The Landau levels are calculated and the corresponding energy scale allows the quantum Hall effect to be realizable at temperatures as high as the room temperature, or even larger than that. Further, the Hall plateaus are observed at certain special integer values and are thus distinct from the scenario for a 2DEG. The discussions are numerically supplemented with the Hall studies for a semi-infinite nanoribbon. Lastly, the experimental demonstration and observation of the quantum Hall effect in graphene are discussed.

Can graphene demonstrate a topological phase? Or does it undergo a topological phase transition as the system parameters are tweaked? Thus, the symmetry aspects, such as the inversion symmetry (parity) and time reversal symmetry, are discussed with a view to explore topological properties of graphene in Chapter 5. The bulk–boundary correspondence and the existence of the edge states in a nanoribbon geometry are investigated, which serve as an acid test for the topological state. Eventually, following Haldane's conjecture, a topological state emerges by breaking the time reversal symmetry where the system acquires a topological gap at the Dirac points. The presence of such a non-trivial gap is confirmed via the presence of chiral edge states in a ribbon. Further, such a system demonstrates the quantum Hall effect without an external magnetic field, which we have computed. Further progress is reported in terms of proposal of a scenario in which the broken time reversal symmetry is repaired using two copies of the Haldane model for each kind of spin of the carriers. This is called the Kane–Mele model, which respects all the symmetries that the parent graphene has. Yet there is an important difference that can be brought about by adding the Rashba spin–orbit coupling, which may be weak, but inherent to the 2D systems. The Rashba term leaves the time reversal symmetry intact. Thus, the Kane–Mele model in the presence of the Rashba spin–orbit contribution yields yet another distinct topological state of matter, namely the quantum spin Hall phase. Owing to the time reversal symmetry being intact, the Chern number vanishes, although the system is characterized by a new topological invariant, known as the $\mathbb{Z}_2$ invariant, which yields the spin Hall conductivity to be non-zero. The prospects of manipulating the spin degree of freedom give birth to an emerging field known as spintronics. The fact that the spin current obeys time reversal symmetry, one gets a non-dissipative transport and thus holds prospects of transmitting information with no (or very little) decay. Thus, the spin transport mechanism does not have any associated Joule heating phenomena.

Finally, in Chapter 6, we have discussed the fractional quantum Hall effect that occurs in cleaner systems where the interparticle interaction becomes dominant and constitutes the only energy scale of the problem. To deal with the overwhelming complexity, Laughlin wrote down a variational wavefunction for the ground state that, other than a ubiquitous Gaussian term, contains a Jastrow factor that keeps two fermions away from each other, thereby enforcing Pauli's exclusion principle. The wavefunction corresponds to a filling fraction of $1/m$, m being an odd integer and denotes the eigenvalue of the z-component of the angular momentum for that state. The Laughlin wavefunction is validated against the exact diagonalization of a system of a few electrons and is found to have a near perfect overlap between the two. One shortcoming of the Laughlin formalism that continued to bother theorists working in the field is the experimental realization of several other values of the fraction (other than $1/m$) at which plateaus are observed. A hierarchy approach is proposed that yields several non-Laughlin fractions encoded via a *parent–daughter* relationship starting with a (*parent*) Laughlin fraction. However, such a scenario suffers from all *daughter* to appear with equal weightage, which is not an experimental reality. Another intuitive solution was proposed by Jain that enunciates an effective medium comprising each fermion trapping an even number of flux quanta. Such a scenario greatly reduces the effective field that the system is subjected to and transforms the fractionally quantized Hall system of interacting fermions to an effectively non-interacting system showing the integer quantum Hall effect. The particle statistics in 2D is discussed with a view to introduce anyons. A brief mention of the abelian and non-abelian anyons is made along with their relevance in studying the FQHE. We supplement the ongoing discussion with a brief mention of the early discoveries of fractionally quantized Hall plateaus in graphene that underscores the role of electron–electron interaction.

Bibliography

[1] N. D. Mermin and H. Wagner. Absence of ferromagnetism or antiferromagnetism in one-or two-dimensional isotropic Heisenberg models. *Phys. Rev. Lett.* **17**, 1133 (1966).

[2] K. S. Novoselov *et al.* Electric field effect in atomically thin carbon films. *Science* **306**, 666 (2004).

[3] K. S. Novoselov *et al.* Two-dimensional gas of massless Dirac fermions in graphene. *Nature* **438**, 197 (2005).

[4] G. Perelman. Grigori Perelman: The genius in hiding. *Science* **314**, 1848 (2006).

[5] D. J. Thouless, M. Kohmoto, M. P. Nightingale and M. den Nijs. Quantized Hall conductance in a two-dimensional periodic potential. *Phys. Rev. Lett.* **49**, 405 (1982).

[6] F. D. M. Haldane. Model for a quantum Hall effect without Landau levels: Condensed-matter realization of the "parity anomaly". *Phys. Rev. Lett.* **61**, 2015 (1988).

[7] C. L. Kane and E. J. Mele. Quantum spin Hall effect in graphene. *Phys. Rev. Lett.* **95**, 226801 (2005); $\mathbb{Z}_2$ topological order and the quantum spin Hall effect. *Phys. Rev. Lett.* **95**, 146802 (2005).

[8] C. Kittel. *Introduction to Solid State Physics*. John Wiley and Sons, USA (1986).

[9] M. V. Berry. Quantal phase factors accompanying adiabatic changes. *Proc. R. Soc. London Ser. A* **392**, 45–57 (1984).

[10] J. Zak. Berry's phase for energy bands in solids. *Phys. Rev. Lett.* **62**, 2747 (1988).

[11] A. Yu. Kitaev. Fault-tolerant quantum computation by anyons. *Annals Phys.* **303**, 2 (2003).

[12] S. Mondal and S. Basu. Topological phase transition induced by band structure modulation in a Chern insulator. *J. Phys. Condens. Mat.* **33**, 225504 (2021).

[13] D. Xiao, M. -C. Chang and Q. Niu. Berry phase effects on electronic properties. *Rev. Mod. Phys.* **82**, 1959 (2010).

[14] T. Fukui, Y. Hatsugai and H. Suzuki. Chern numbers in discretized Brillouin zone: Efficient method of computing (spin) Hall conductances. *J. Phys. Soc. Jpn.* **74**, 1674 (2005).

[15] Yu. A. Bychkov and E. I. Rashba. Properties of a 2D electron gas with lifted spectral degeneracy. *Sov. Phys. -JETP Lett.* **39**, 78 (1984).

[16] A. Altland and M. R. Zirnbauer. Nonstandard symmetry classes in mesoscopic normal-superconducting hybrid structures. *Phys. Rev. B* **55**, 1142 (1997).

[17] S. Ryu *et al.* Topological insulators and superconductors: Tenfold way and dimensional hierarchy. *New J. Phys.* **12**, 065010 (2010).

[18] L. Fu and C. L. Kane. Time reversal polarization and a $\mathbb{Z}_2$ adiabatic spin pump. *Phys. Rev. B* **74**, 195312 (2006).

[19] R. J. Triebl. *Wannier charge centers and the calculation of topological invariants: Application to the Kane-Mele-Hubbard model* (Unpublished Master Thesis). Graz University of Technology (2015).

[20] H. Min, J. E. Hill, N. A. Sinitsyn, B. R. Saha, L. Kleinman and A. H. McDonald. Intrinsic and Rashba spin–orbit interactions in graphene sheets. *Phys. Rev. B* **74**, 165310 (2006).

[21] B. A. Bernevig, T. L. Hughes and S. C. Zhang. Quantum spin Hall effect and topological phase transition in HgTe quantum wells. *Science* **314**, 1757 (2006).

[22] M. König *et al.* Quantum spin hall insulator state in HgTe quantum wells. *Science* **318**, 766 (2007).

[23] For a brief review of Kerr effect, see Q. Zhong and J. T. Forukas. Optical Kerr effect spectroscopy of simple liquids. *J. Phys. Chem. B* **112**, 15529 (2008).

[24] C. Stamm *et al.* Magneto-optical detection of the spin Hall effect in Pt and W thin films. *Phys. Rev. Lett.* **119**, 087203 (2017).

[25] Y. K. Kato, R. C. Myers, A. C. Gossard and D. D. Awschalom. Current-induced spin polarization in strained semiconductors. *Phys. Rev. Lett.* **93**, 176601 (2004).

[26] J. Wunderlich, B. Kaestner, J. Sinova and T. Jungwirth. Experimental observation of the spin-Hall effect in a two-dimensional spin–orbit coupled semiconductor system. *Phys. Rev. Lett.* **94**, 047204 (2005).

[27] J. Sinova, S. O. Valenzuela, J. Wunderlich, C. H. Back and T. Jungwirth. Spin Hall effects. *Rev. Mod. Phys.* **87**, 1213 (2015).

[28] Q.-F. Sun and X. C. Xie. Definition of the spin current: The angular spin current and its physical consequences. *Phys. Rev. B* **72**, 245305 (2005).

[29] G. P. Fisher. Definition of the spin current: The electric dipole moment of a moving magnetic dipole. *Am. J. Phys.* **39**, 1528 (1971).

[30] S.-Q. Shen. Spintronics and spin current. *AAPPS Bulletin* **18**, 29 (2008).

[31] E. Majorana. Teoria simmetrica dell'elettrone e del positrone. *Nuovo Cimento* **5**, 171 (1937).

[32] L. Fu and C. L. Kane. Superconducting proximity effect and Majorana fermions at the surface of a topological insulator. *Phys. Rev. Lett.* **100**, 096407 (2008).

[33] L. Fu and C. L. Kane. Josephson current and noise at a superconductor/quantum-spin-Hall-insulator/superconductor junction. *Phys. Rev. B* **79**, 161408(R) (2009).

[34] R. Lutchyn, J. D. Sau and S. Das Sarma. Majorana fermions and a topological phase transition in semiconductor-superconductor heterostructures. *Phys. Rev. Lett.* **105**, 077001 (2010).

[35] Y. Oreg, G. Rafael and F. V. Oppen. Helical liquids and Majorana bound states in quantum wires. *Phys. Rev. Lett.* **105**, 17002 (2010).

[36] J. Nilsson, A. R. Akhmerov and C. W. Beenakker. Splitting of a Cooper pair by a pair of Majorana bound states. *Phys. Rev. Lett.* **101**, 120403 (2008).

[37] K. T. Law, P. A. Lee and T. K. Ng. Majorana fermion induced resonant Andreev reflection. *Phys. Rev. Lett.* **103**, 237001 (2009).

[38] M. Creutz. End states, ladder compounds, and domain-wall fermions. *Phys. Rev. Lett.* **83**, 2636 (1999).

[39] Y. Kuno, T. Orito and I. Ichinose. Flat-band many-body localization and ergodicity breaking in the Creutz ladder. *New J. Phys.* **22**, 013032 (2020).

[40] A. Altland and M. R. Zirnbauer. Nonstandard symmetry classes in mesoscopic normal-superconducting hybrid structures. *Phys. Rev. B* **55**, 1142 (1997).

[41] J. Zurita *et al.* Topology and interactions in the photonic Creutz and Creutz-Hubbard ladders. *Adv. Quantum Technol.* **3**, 1900105 (2020).

[42] S. Dutta. *Electronic Transport in Mesoscopic Systems*. Cambridge University Press, Cambridge (1995).

[43] Y. Imry. *Electronic Transport in Mesoscopic Systems*. Oxford University Press, New York (1997).

[44] K. Hess. Boltzmann transport equation. In *The Physics of Submicron Semiconductor Devices*, ed. H. L. Grubin *et al.* Springer New York, NY (1988).

[45] R. Landauer. Spatial variation of currents and fields due to localized scatterers in metallic conduction. *IBM Journal of Research and Development* **1**, 223 (1957).

[46] P. A. Mello and N. Kumar. *Quantum Transport in Mesoscopic Systems: Complexity and Statistical Fluctuations*. Oxford University Press, New York (2004).

[47] Ando, Y. Arakawa, K. Furuya, S. Komiyama and H. Nakashima, eds. *Mesoscopic Physics and Electronics*. Springer, New York (1998).

[48] Y. Aharonov and D. Bohm. Significance of electromagnetic potentials in the quantum theory. *Phys. Rev.* **115**, 485 (1959).

[49] B. J. van Wees, H. van Houten, C. W. J. Beenakker, J. G. Williamson, L. P. Kouwen-hoven, D. van der Marel and C. T. Foxon. Quantized conductance of point contacts in a two-dimensional electron gas. *Phys. Rev. Lett.* **60**, 848 (1988).

[50] D. A. Wharam *et al.* One-dimensional transport and the quantisation of the ballistic resistance. *J. Phys. C: Solid State Physics* **21**, L209 (1988).

[51] G. Bergmann. Weak localization in thin films: A time-of-flight experiment with conduction electrons. *Phys. Reports* **107**, 1 (1984).

[52] P. A. Lee and T. V. Ramakrishnan. Disordered electronic systems. *Rev. Mod. Phys.* **57**, 287 (1985).

[53] R. G. Garcia. Atomic-scale manipulation in air with the scanning tunneling microscope. *Applied Phys. Lett.* **60**, 1960 (1992).

[54] J. M. Leinaas and M. J. Myrheim. On the theory of identical particles. *Il Nuovo Cimento* **37B**, N.1 (1977).

[55] F. Wilczek. Quantum mechanics of fractional-spin particles. *Phys. Rev. Lett.* **49**, 957 (1982).

[56] X. Ying, V. Bayot, M. B. Santos and M. Shaygon. Observation of composite-fermion thermopower at half-filled Landau levels. *Phys. Rev. B* **50**, 4969 (1994).

[57] U. Zeitlar *et al.* Investigation of the electron-phonon interaction in the fractional quantum Hall regime using the thermoelectric effect. *Phys. Rev. B* **47**, 16008(R) (1993).

[58] D. R. Leadley, R. J. Nicholas, C. T. Foxon and J. J. Harris. Measurements of the effective mass and scattering times of composite fermions from magnetotransport analysis. *Phys. Rev. Lett.* **72**, 1906 (1994).

[59] B. I. Halperin, P. A. Lee and N. Read. Theory of the half-filled Landau level. *Phys. Rev. B* **47**, 7312 (1993).

[60]   K. V. Klitzing, G. Dorda and M. Pepper. New method for high-accuracy determination of the fine-structure constant based on quantized Hall resistance. *Phys. Rev. Lett.* **45**, 494 (1980); K. V. Klitzing, 25 Years of Quantum Hall Effect (QHE) A Personal View on the Discovery, Physics and Applications of this Quantum Effect. *Séminaire Poincaré* **2**, 1 (2004).

[61]   G. Landwehr. The discovery of the quantum hall effect. *Metrologia* **22**, 118 (1986).

[62]   D. C. Tsui, H. L. Stormer and A. C. Gossard. Two-dimensional magnetotransport in the extreme quantum limit. *Phys. Rev. Lett.* **48**, 1559 (1982).

[63]   C. Kittel. *Introduction to Solid State Physics* (8th ed.). John Wiley and Sons, USA (2004).

[64]   N. W. Ashcroft and N. D. Mermin. *Solid State Physics*. Harcourt College Publishers, USA (1976).

[65]   L. Onsager. Reciprocal relations in irreversible processes. I. *Phys. Rev.* **37**, 405 (1931).

[66]   H. Goldstein, C. P. Poole and J. Safko. *Classical Mechanics* (3rd ed.). Addison Wesley, USA (2002).

[67]   R. Shankar. *Principles of Quantum Mechanics* (2nd ed.). Kluwer Academic/Plenum Publishers, New York, USA (1994).

[68]   G. D. Mahan. *Many Particle Physics*. Springer, Springer New York, NY (3rd Ed.) (2000).

[69]   D. Tong. Lectures on the quantum Hall effect. arXiv:1606.06687 (2016).

[70]   D. J. Thouless, M. Kohmoto, M. P. Nightingale and M. den Nijs. Quantized Hall conductance in a two-dimensional periodic potential. *Phys. Rev. Lett.* **49**, 405 (1982).

[71]   K. S. Novoselov *et al*. Room-temperature quantum Hall effect in graphene. *Science* **315**, 1379 (2007).

[72]   R. S. Deacon *et al*. Cyclotron resonance study of the electron and hole velocity in graphene monolayers. *Phys. Rev. B* **76**, 081406(R) (2007).

[73]   G. Li, A. Luican and E. Y. Andrei. Scanning tunneling spectroscopy of graphene on graphite. *Phys. Rev. Lett.* **102**, 176804 (2009).

[74]   E. McCann and V. I. Fálko. Landau-level degeneracy and quantum Hall effect in a graphite bilayer. *Phys. Rev. Lett.* **96**, 086805 (2006).

[75]   R. Bistritzer and A. H. McDonald. Moiré bands in twisted double-layer graphene. *Proc. Natl. Acad. Sci. USA* **108**, 12233 (2011).

[76]   G. Wagner *et al*. Global phase diagram of the normal state of twisted bilayer graphene. *Phys. Rev. Lett.* **128**, 156401 (2022).

[77]   V. I. Fálko. Electronic properties and the quantum Hall effect in bilayer graphene. *Phil. Trans. R. Soc.* **366**, 205 (2008).

[78]   J. Nilsson, A. H. Castro Neto, N. M. R. Peres and F. Guinea. Electron–electron interactions and the phase diagram of a graphene bilayer. *Phys. Rev. B* **73**, 214418 (2006).

[79]   E. Macaan, V. I. Fálko and Y. Zhang. Landau-level degeneracy and quantum Hall effect in a graphite bilayer. *Phys. Rev. Lett.* **96**, 086805 (2006).

[80]   G. Li and E. Andrei. Observation of Landau levels of Dirac fermions in graphite. *Nature Phys.* **3**, 623 (2007).

[81]   D. S. Lee *et al*. Quantum Hall effect in twisted bilayer graphene. *Phys. Rev. Lett.* **107**, 216602 (2011).

[82] Z. H. Ni *et al.* Reduction of Fermi velocity in folded graphene observed by resonance Raman spectroscopy. *Phys. Rev. B* **77**, 235403 (2008).

[83] Y. Kim *et al.* Odd integer quantum Hall states with interlayer coherence in twisted bilayer graphene. *Nano Lett.* **21**, 4249 (2021).

[84] M. Sprinkle *et al.* First direct observation of a nearly ideal graphene band structure. *Phys. Rev. Lett.* **103**, 226803 (2009).

[85] K. Wakabayashi, Y. Takane, M. Yamamoto and M. Sigrist. Electronic transport properties of graphene nanoribbons. *New J. Phys.* **11**, 095016 (2009).

[86] D. R. Hofstadter. Energy levels and wave functions of Bloch electrons in rational and irrational magnetic fields. *Phys. Rev. B* **14**, 2239 (1976).

[87] M. Ya. Azbel. Energy spectrum of a conduction electron in a magnetic field. *JETP* **19**, 634 (1964).

[88] T. Schlösser, K. Ensslin, J. P. Kotthaus and M. Holland. Landau subbands generated by a lateral electrostatic superlattice-chasing the Hofstadter butterfly. *Semicond. Sci. Technol.* **11**, 1582 (1996).

[89] A. H. Castro Neto, F. Guinea, N. M. R. Peres, K. S. Novoselov and A. K. Geim. The electronic properties of graphene. *Rev. Mod. Phys.* **81**, 109 (2009).

[90] P. G. Harper. Single Band Motion of Conduction Electrons in a Uniform Magnetic Field. *Proc. Phys. Soc. A* **68**, 874 (1955).

[91] R. Rammal. Landau level spectrum of Bloch electrons in a honeycomb lattice. *Journal de Physique* **46**, 1345 (1985).

[92] Jain, Jainendra K. *Composite Fermions.* Cambridge University Press, Cambridge (2007).

[93] R. B. Laughlin. Anomalous quantum Hall effect: An incompressible quantum fluid with fractionally charged excitations. *Phys. Rev. Lett.* **50**, 1395 (1983).

[94] F. D. M. Haldane. Fractional quantization of the Hall effect: A hierarchy of incompressible quantum fluid states. *Phys. Rev. Lett.* **51**, 605 (1983).

[95] B. I. Halperin. Statistics of quasiparticles and the hierarchy of fractional quantized Hall states. *Phys. Rev. Lett.* **52**, 1583 (1984).

[96] B. I. Halperin. Statistics of quasiparticles and the hierarchy of fractional quantized Hall states. *Phys. Rev. Lett.* **52**, 2390 (1984).

[97] D. Glattli *et al.* Shot noise and the Luttinger liquid-like properties of the FQHE. *Physica E* **6**, 22 (2000).

[98] J. M. Leinaas and J. Myrheim. On the theory of identical particles. *Il Nuovo Cimento* **37B**, N.1 (1977).

[99] F. Wilczek. Quantum mechanics of fractional-spin particles. *Phys. Rev. Lett.* **49**, 957 (1982).

[100] A. Kitaev. *Proc. of Symp. in Applied Math* (2000); J. Preskill, Lecture Notes for Physics 219, California Institute of Technology (2004).

[101] R. S. Chen. Generalized Yang–Baxter equations and braiding quantum gates. *Journal of Knot Theory and Its Ramifications* **21**, 1250087 (2011).

[102] R. L. Willet *et al.* Observation of an even-denominator quantum number in the fractional quantum Hall effect. *Phys. Rev. Lett.* **59**, 1776 (1987).

[103] X. Du *et al.* Fractional quantum Hall effect and insulating phase of Dirac electrons in graphene. *Nature* **462**, 192 (2009).

[104] K. I. Bolotin *et al.* Observation of the fractional quantum Hall effect in graphene. *Nature* **462**, 196 (2009).

[105] C. R. Dean *et al.* Multicomponent fractional quantum Hall effect in graphene. *Nature Phys.* **7**, 693 (2011).

[106] L. Wang *et al.* Evidence for a fractional fractal quantum Hall effect in graphene superlattices. *Science* **350**, 1231 (2015).

[107] Z. Jian *et al.* Quantum Hall states near the charge-neutral dirac point in graphene. *Phys. Rev. Lett.* **99**, 106802 (2007).

[108] Y. Zhang *et al.* Landau-level splitting in graphene in high magnetic fields. *Phys. Rev. Lett.* **96**, 136806 (2006).

[109] K. Nomura and A. H. McDonald. Quantum Hall ferromagnetism in graphene. *Phys. Rev. Lett.* **96**, 256602 (2006).

[110] J. Alicea and M. P. A. Fisher. Graphene integer quantum Hall effect in the ferromagnetic and paramagnetic regimes. *Phys. Rev. B* **74**, 075422 (2006).

[111] K. Yang. Spontaneous symmetry breaking and quantum Hall effect in graphene. *Solid State Commun.* **143**, 27 (2007).

[112] E. V. Gorbar *et al.* Dynamics in the quantum Hall effect and the phase diagram of graphene. *Phys. Rev. B* **78**, 085437 (2008).

[113] B. E. Feldman *et al.* Unconventional sequence of fractional quantum Hall states in suspended graphene. *Science* **337**, 1196 (2012).

Index